SpringerBriefs in Ethics

Springer Briefs in Ethics envisions a series of short publications in areas such as business ethics, bioethics, science and engineering ethics, food and agricultural ethics, environmental ethics, human rights and the like. The intention is to present concise summaries of cutting-edge research and practical applications across a wide spectrum. *Springer Briefs in Ethics* are seen as complementing monographs and journal articles with compact volumes of 50 to 125 pages, covering a wide range of content from professional to academic. Typical topics might include:

- Timely reports on state-of-the art analytical techniques
- A bridge between new research results, as published in journal articles, and a contextual literature review
- A snapshot of a hot or emerging topic
- In-depth case studies or clinical examples
- Presentations of core concepts that students must understand in order to make independent contributions

More information about this series at http://www.springer.com/series/10184

John Forge

The Morality of Weapons Research

Why it is Wrong to Design Weapons

 Springer

John Forge
School of History and Philosophy of Science
The University of Sydney
Sydney, NSW, Australia

ISSN 2211-8101   ISSN 2211-811X   (electronic)
SpringerBriefs in Ethics
ISBN 978-3-030-16859-9   ISBN 978-3-030-16860-5   (eBook)
https://doi.org/10.1007/978-3-030-16860-5

Library of Congress Control Number: 2019935993

This Springer imprint is published by the registered company Springer Nature Switzerland AG
The registered company address is: Gewerbestrasse 11, 6330 Cham, Switzerland

For my grandson Sasha Barretto

Preface

My interest in the topic of this book, the morality of weapons research, goes back some 25 years to the time I was teaching a course on science and ethics in a Department of Science and Technology Studies (STS). The Manhattan Project, of which more about in this book, had been standardly used as a case study in STS in regard to the relations between science and government. The use of the atomic bombs was also a topic for discussion. My interest was with the responsibility of the scientists who designed the bombs for their use, and when examining that issue it becomes clear that there would have been no atomic weapons available by the end of the war had not a number of scientists managed to convince the governments of Britain and the United States to sponsor the research. However, the scientists in question were worried about having a means to deter Germany should it make an atomic bomb, not with destroying Japanese cities. Japan was not in the war when the first research project was set up in Britain and many of these pioneers regretted what they did. I began to think about this issue, about weapons research being done in one context for a given reason and then being used in another. I wondered whether this was true in general, for all weapons research. What is true is that this is certainly possible, because what weapons research does is to create designs and these survive the beginnings and ends of war, changes of governments, changes of international relations and so forth. I eventually came to the conclusion that undertaking weapons research is always morally wrong and always morally unjustifiable.

My first monograph on this subject, *Designed to Kill: The Case Against Weapons Research*, was published by Springer in 2012. So I am very pleased that the same publisher is producing this much shorter, Springerbrief, version of my argument against weapons research. I should add that while this is essentially the same argument as before, it is couched in quite different terms to that of my first book—it is not a summary or précis of that work. So I would like to express my thanks to Floor Oosting, Christopher Wilby, who was one of my previous editors, and all the other excellent people at Springer. I'm also grateful to the referees who made interesting and helpful suggestions to my original proposal.

Sydney, Australia   John Forge

Contents

Introduction

The aim of this book is to 'make the case against weapons research', that is, to provide good reasons to accept that weapons research is always morally wrong and never morally justified. I will briefly state my argument in favour of these claims, make some comments by way of clarification, comment and interpretation, and give a quick sketch of the chapters which follow.

Outline of the Case Against Weapons Research

To begin, we can all agree that if one person harms another, then this is *prima facie* morally wrong. The force of '*prima facie*' here is that if all we know is that an action which harms someone has taken place, then we immediately judge that action to be (morally) wrong and consider the person who did it blameworthy until further notice. The judgement may be withdrawn or qualified when more is known about the circumstances in which the harming took place. For example, if the person who was charged with harming in fact did not do so—it was someone else, or the action did not actually harm anyone, for example—then she has an *excuse*: she did not do what she was accused of doing. When someone has an excuse, they are not to be blamed for what they did, and we withdraw the judgement that the action was morally wrong. Another possibility is that the person in question deliberately and intentionally caused the harm, but did so because she was defending herself. Provided that the harm she inflicted was not too excessive, then under these circumstances, she is also not to be blamed because her action was *justified*. In this case, we also normally withdraw the original judgement that her action was morally wrong, though I prefer to say in such a case that it turned out to be morally *permissible*, as this acknowledges that harm was deliberately caused. This leaves those instances where there is no excuse or justification and these are such that the original judgement is re-affirmed and the person who did the harming is held to be morally blameworthy.

Talking about harming and moral wrongdoing is relevant here because weapons are the means to harm. Weapons are specifically designed to harm, and they are the most effective means available for harming. This is not controversial. The standard justifications, what I call as a whole the *standard rationale*, for everything to do with weapons, weapons acquisition, weapons production and so on, including weapons research, research that aims to design weapons, are defence and deterrence. The assumption is that weapons are the best way to defend oneself from others, who also have weapons, and, better still, deter them from using their weapons, because weapons harm. I will argue that defence and deterrence are what I call *derivative* functions of weapons, these uses of weapons 'derive from' their *primary* function, which is as means to harm. This claim and its implications, it turns out, are far from uncontroversial. I go on to argue that if harming is wrong, then so is providing the means to harm to others. I call this the *means principle*. And it follows that weapons research is morally wrong. This, in terse outline, is the first step in my case against weapons research. However, we have seen that judgements about the moral wrongdoing may be withdrawn if the person who performed the action had an excuse, and moreover, such actions may turn out to be morally permissible if the agent had justification.

According to the standard rationale, weapons research is carried out for defence, or for deterrence (or both). So it seems that such defensive or deterrent weapons research aims to produce defensive or deterrent weapons. If it were true that there are weapons that could only defend or only deter, weapons that could only cause harm if employed to prevent harm, then would not a weapons researcher who undertook to design such weapons have an excuse? Could she not say that she was not producing the means to harm but, primarily, the means to defend against harm? This response amounts to an excuse because the agent did not do what she was accused of doing, she did not produce the means to harm. On my account of the matter, this response implies that weapons can have more that one primary purpose: they can be the means to harm, the means to defend against harm and possibility also the means to deter harm. In my view of designer responsibility, what designers need to answer for or are called to account over, is the primary purpose of what it is they design. They are, in my view, 'committed to' the primary purpose. I reply to these claims about defence and deterrence by showing that there are no weapons that are either defensive or deterrent in the requisite sense, and hence establish that weapons are indeed primarily the means to harm.

I have considered the claim that weapons research intended to design defensive or deterrent weapons as an excuse, which seeks to have the charge of moral wrongdoing dropped altogether. However, defence and deterrence can also be put forward as justifications. Suppose it is accepted that weapons are primarily the means to harm, and as such, they can be used to defend against harm by harming aggressors, those who initiate hostilities. Justification aims to establish that the harmful act in question was morally permissible because it prevented further harms. An important distinction here is this: if there actually were weapons that could only prevent harm, then it would not be necessary to refer to the circumstances or context in which the corresponding weapons research was carried out. The actual historical circumstances would not be relevant because the only 'missions' the weapons could take part in would be to prevent harm. The only thing that would be relevant would be the technical character of the weapons which render them incapable of doing anything but preventing harm. The attempted denial of moral wrongdoing here is what I call *ahistorical* because there is no reference to the context of the weapons research. By contrast, justifications that accept that weapons are the means to harm but that they are needed to stop others from harming must refer to the context in which the work is to be conducted, to the actual historical circumstances in which harm is prevented. Such justifications are therefore *historical*. But the problem with such historical justification is this: how can anyone, the weapons researcher included, know that the products of her work will prevent more harm than they cause?

Weapons research produces *knowledge*, in the form of designs. This knowledge enables those with the requisite skills and materials to make the weapons in question, the only limits on weapons manufacture being the materials, skills, economic costs and desire to have the weapons. The designs themselves do not wear out or impose any limits on how many times they are reproduced, nor on the time or place where they are used. Designs project into the future, and they tend to spread out. To take an example, which I will discuss in more detail in the book, the first nuclear weapons were designed during the Manhattan Project, between 1942 and 1945, and three nuclear weapons, atomic bombs, were produced. There were only three made because of the limited amount of (fissile) material. There are other kinds of nuclear weapons available now, much more powerful thermonuclear weapons, which all incorporate systems based on one of the two original designs as 'fission triggers', primary components which ignite the secondary. The US still keeps the original Manhattan Project designs under tight control. However, essentially the same designs were discovered, with only slight variations, by the Soviet Union, Britain, France and China, and Israel, India, Pakistan, South Africa, and one assumes Iran and North Korea. All of these states, with the probable exception of Iran, have made nuclear weapons very similar to the three originals and now the majority also have thermonuclear weapons.

No one knows if nuclear weapons will be used again. Most certainly the people who worked on the Manhattan Project, all deceased, could not know this, indeed, they did not know, until the very end, that their weapons would be used to kill Japanese civilians in Hiroshima and Nagasaki. But their work brought nuclear weapons into existence and set up the US nuclear weapons programme, which in turn inspired the Soviet Union and others to follow suit. The Manhattan Project weapons researchers could not have known this either. Weapons researchers are simply not in a position to know what they must know before they undertake their work if this is not to judged morally wrong and for them to be held blameworthy. Any attempted justification of weapons research will have great difficulty even getting past the first step of determining what harms the weapons will cause, let alone working out how much harms would thereby be prevented, and then estimating that the latter outweigh the former. I conclude that not only is weapons research morally wrong, it is also morally unjustifiable.

I think many people may be uneasy about the pace of weapons acquisition and the kind of weapons that have been procured, nuclear weapons in particular. I think many believe that producing more and more weapons makes the world less, not more, safe. But perhaps they are seduced by the mantra of defence and believe that weapons are a necessary evil. If so, I hope my argument will make my readers revise their opinions, or confirm their intuitions, and agree with me that weapons research should cease because it is morally wrong. However, even if the case against weapons research can be made, the prospect of the activity coming to a halt is remote, to say the least. In which case, it may seem that this book is simply an interesting—if nothing else, it is interesting—exercise in applied philosophy, and nothing more. This response raises a number of questions and issues about the point or aim of doing philosophy, philosophy in general, and applied ethics in particular. I have three comments.

I would be surprised if the scientists, engineers and other specialists who are engaged in weapons research do not accept the standard rationale for their work and believe that what they are doing is providing the means to prevent harm; or at lease I believe that this would be how they would represent to themselves what they do if they reflected on it at all. So my first comment is that if (some) weapons researchers knew that what they do is morally wrong, maybe they would do something else. If that is true, then it is a good idea to try to inform them. As a second comment, I note that activities that were once done routinely are now proscribed and illegal in many countries because it came to be accepted that they are morally wrong—slavery is an obvious example (Forge 2018: x–xii). There can therefore be *moral progress*, but for this to be possible, it needs to be demonstrated and accepted that the activities in question are morally wrong. Finally, although many people and many institutions are not themselves moved by the demands of morality, nevertheless they realise that others are and they realise that it is not good for them to be seen to do what is morally wrong. Therefore, they have reason to respond and try to justify themselves, even if they are not sincere. What follows therefore need not be understood as a purely intellectual exercise: if I am able to show that people ought not to undertake weapons research, maybe they will eventually stop doing so?

Turning to two other matters, I am going to understand a weapon to be a military weapon or something intended for the military, weapons for warfare. Thus, research aimed at producing more effective means for law enforcement, or for recreation, is not something I will address. It is true that some 'military-style' weapons are used by police and other law-enforcement agencies, especially riot police and other heavily armed paramilitaries. But it seems that these weapons are the product, or by-product, of military weapons research. If not, then again I will have nothing directly to say about the research in question, though I think it will be clear how to extend my argument in that direction. In the second place, I am not going to discuss non-lethal weapons. I have discussed these elsewhere and again I think it will be clear how my argument applies to weapons research aimed to produce weapons of this kind (Forge 2012: 173–175).

Outline of the Book

The book has six chapters. The first chapter develops the framework that will be used to make moral judgements about weapons research. This is a 'bare bones' morality which maintains that it is wrong to harm without justification—and so is entirely uncontroversial. Given this principle, I contend, along with Bernard Gert, that the only justification for harming is if the harmful act prevents at least as much harm as it causes. The standard rationale for everything to do with weapons and war is defence, the protection of people and their assets, and hence Gert's justificationary principle is appropriate to the present inquiry. In Chap. 2, I discuss two examples of weapons research in some detail. The first of these, the Manhattan Project, shows how scientific theory can guide weapons research, and the second, the evolution of torsion artillery, shows that weapons research can be conducted without knowledge of the underlying scientific principles. I conclude from this comparison that weapons research has been done for at least two and a half thousand years. In this chapter, I describe weapons research as the activity aimed to design new weapons systems or improve existing systems.

In Chap. 3, I give reasons why weapons should be taken to be the means to harm and that this is the 'canonical description'. Thus, weapons are not primarily the means to defend or the means the deter, but only have these roles because they are the means to harm. The primary aim of weapons research is therefore to design new ways to harm. I argue that it is morally wrong to do this, which is the first step in my case against weapons research. In Chap. 4 I talk about defence and address the issue as to whether there are inherently defensive weapons, weapons that cannot aid aggression in any way, and hence whether the conclusion of Chap. 3 needs to be revised and weapons research aimed to develop such weapons allowed as an exception to the rule. I maintain that there are no such weapons and, moreover, aggressive wars of conquest are punctuated by periods of defence and re-grouping, so weapons that are best suited to defensive roles are needed for that purpose.

Chapters 5 and 6 are concerned with justification. I argue that weapons research can never be justified because the demand that more harms are seen to be prevented than caused cannot be satisfied. Indeed, the harms caused cannot be determined, let alone the harms prevented and a comparison made between the two 'amounts' of harm. In Chap. 5, I discuss two possible ways of side-stepping this problem. The first holds that weapons research done by democratic states is justifiable because such states do not embark of aggressive wars, and hence, any weapons they possess will only be used for defence, so the prevention of harm. However, it is simply false that democratic states are never the aggressor. The second suggestion is that weapons research done to aid a just war is acceptable. However, I show that weapons research is in fact incompatible with the theory of just war. So the task of estimating the harms caused and harms prevented, and then comparing them remains. In Chap. 6, I give three examples to show that there is no prospect of this being possible. I begin Chap. 6 by talking about the idea of the context of weapons research and identify two 'contextual factors' that condition weapons research. One of the aims of the examples is to show what happens when contexts change. I then review and summarise the argument in my conclusion. I have included five appendices. Unlike the six main chapters, these are not concerned with the development of the main argument, but rather elaborate matters raised in the chapters (appendices 1 and 4) or address certain implications of the main argument as these arise (appendices 2, 3 and 5).

References

Forge, J. 2012. *Designed to Kill: The Case Against Weapons Research*. Dordrecht: Springer.
Forge, J. 2018. *The Morality of Weapons Design and Development*. IGI: Hershey, PA.

Chapter 1
Morality and the Harm Principle

In order to argue that weapons research is morally wrong and, in addition, that it is morally unjustifiable, it is necessary to appeal to some system of morality. This is because it is necessary, in the first place, to understand just exactly what is at stake. When I claim that weapons research is morally wrong, what am I saying, and what does it mean to say that weapons research is morally wrong? In the second place, it is necessary to appeal to a system of morality for justification. Why is it that weapons research is morally wrong? And finally it is necessary to appeal to a system of morality to show that weapons research is not justifiable. The distinction between the two claims I hope to establish in this book, that weapons research is morally wrong and that it is morally unjustifiable, is to some extent a feature of the particular moral system that I favour—as we will see, not all moral systems embody such a distinction. This raises the question as to the possibility of alternative moral systems and hence different judgements about the morality of weapons research. There are moral systems that are different from the one I am going to use but I am confident that they do not do so in regards to judgements about weapons research. All moral systems incorporate the harm principle in one form or another. A feature of the system I adopt, which is a pared down or 'bare bones' type morality, is that it comprises little more that the harm principle and its implications.[1] I will show how alternative moral systems can incorporate the principle in the second section of this chapter.

---

[1]In my other books on weapons research, Forge (2012, 2018), I have identified common morality with the moral system advocated by Bernard Gert—see Gert (2005) for a comprehensive statement and Gert (2004) for an accessible introduction. Here I approach common morality in a slightly different way, but I still use and endorse Gert's account. The approach is intended to facilitate the discussion of weapons research, and is tailored to that end.

J. Forge, *The Morality of Weapons Research*, SpringerBriefs in Ethics,
https://doi.org/10.1007/978-3-030-16860-5_1

1.1   The Harm Principle

Morality, as with the law, limits of the free conduct and agency of the individual, and comparing moral strictures with those of the law is a useful way to begin. A central question in political philosophy, especially in the Liberal tradition, is to what extent the state, though the offices of the law, should restrict the freedom of individuals who make up civil society. The answer given by John Stuart Mill appeals to what has become known as the Harm Principle: "That the only purpose for which power can be rightfully exercised over any member of a civilized community, against his will, is to prevent harm to others." (Mill 1954a: 73). While we can all agree that the preventing harm is an important, even an overriding concern of the state, Mill's statement of the Harm Principle prompts a number of further questions. For instance, how does the state 'rightfully exercise' power, how, in practice, are laws to be formulated and enforced, and what penalties should be exacted on those who disobey them? These questions presuppose that we know what harm, or harms, are, for there can be no legislating against harm if one does not know what it is. This issue, unlike those concerned with actual legislation, marks ground shared between morality and politics: morality is concerned with the prevention of harm, but without enforcement of conformity by state power. The question naturally arises as to just what it is that should make people limit their behaviour, actions, choices, decisions, etc., in order to avoid harming others. I will come back to this matter when we have come to some decisions about what harming and harms are.

One way to state what harm is is simply to give a list of all those acts that count as harming. This would be time-consuming, generate a long list and run the risk of omitting things that should be included. Another approach is to categorise harms into types of act. This is essentially what Bernard Gert does (Gert 2005: 20–21). His moral system comprises ten moral rules, the first eight of which prohibit a particular kind of harming. Here are his first two rules:

> Rule 1: Do not kill.
>
> Rule 2: Do not cause pain.

Killing and causing pain are clearly harms, and acts which kill and cause pain are clearly examples of harming—indeed, these are paradigm cases of harms and harming: in the normal course of events, *no one ever* wants to be killed or be in pain. In my previous work on weapons research I adopted a version of Gert's moral system and made use of the rules that specifically mentioned kinds of harms, such as Rules 1 and 2 (Gert 2005: 112). This is because it seems that the kind of harms caused by weapons research are most obviously physical injuries and death. Here, however, I am going to take a slightly different tack, and ask whether we can give a general account of harm, such that what are mentioned in Gert's rules are particular kinds of harming. This will help us have a *general* understanding of what harm is.[2]

Joel Feinberg has considered this matter in his study of the moral basis of criminal law (in Feinberg 1984). Feinberg begins with Mill's statement of the Harm Principle and then asks, as we did, just what is meant by "harm". After some careful discussion, he says "only setbacks to interests that are wrongs …are to count as harms in the appropriate sense" (Feinberg 1984: 36). As to interests

> …interests …are distinguishable components of a person's well being: he flourishes or languishes as they flourish or languish. What promotes them is to his advantage, or *in his interest*; what thwarts them is to his detriment, or *against his interest*. They can be blocked or defeated by events of an impersonal nature or by plain bad luck. But they can only be 'invaded' by human beings…It is only when an interest is thwarted through an invasion [by human beings] that its possessor is harmed in the legal sense. One person harms another … by invading or setting back his interest. (Feinberg 1984: 34, original emphasis)

In this passage Feinberg says that the setting back of an interest of one person by another is harming in the legal sense; I will assume here that this is also true of harming in the moral sense.

On this basis, we can come up with following rule based on Feinberg's interpretation of the harm principle:

> Do not invade, and so set back, the interests of others.

One should not invade the interest of another because that is to harm them, and moral persons should voluntarily limit their freedom to act by foregoing any action that would harm another person (or moral subject).[3] So what we have from Feinberg is a single sense of harm, namely the setting back of an interest, which lets us formulate the principle. But does this really help? We began this section by stating Mill's Harm Principle as the starting point for developing a moral system, looked for a way to characterise harm and harming, and now we have an account thereof in terms of interests. Does this not substitute one poorly understood idea for another one, one which seems somewhat abstract and artificial? Feinberg has however provided some helpful clarification: interests are about all aspects of our *well-being*, we flourish or do well when they flourish, we do badly and languish when they do. The kinds of harms Gert refers to in the two rules are clearly set-backs to interests. Things go badly for us when we are in pain, and that is certainly true if we are killed. The other rules of Gert's system mention being disabled, lied to, cheated, deprived of freedom and pleasure, and clearly these too are set-backs to interests. Interests are what are important to us, essential even, and

---

[2]I find Gert's system very appealing, which might perhaps have led to my not recognising that others might want more to be said about what harm is. I take the opportunity to do so here. However, the following remarks about Feinberg's views are intended to underpin Gert's system, not replace it.

[3]Many, myself included, think animals as well as humans should be 'protected' by morality, even though they are not moral agents.

as Feinberg says, we flourish when they do, and we do poorly when they do not. So in fact we have not substituted one poorly understood notion for another, but rather have provided a unified idea of the locus of harm: our interests are what can be harmed, or rather we are harmed when our interests are set-back.

The moral principle just formulated should not, however, be obeyed in all circumstances, come what may. Consider this example: would not aiding a runaway slave be a set-back to the interest of another, namely the legal owner of the slave, and as such is it not proscribed by the principle; and if that is the case, is there not something wrong with the principle itself? Slave owning was certainly an interest of those who lived in the South before the Civil War, as it has been to most slave owners in history. Slaves do unpaid work and hence can be used for economic benefit. What this example shows is that the principle formulated above may be sufficient to fix the class of harms, but it is not yet a *moral* principle, for it cannot be morally permissible to support a practice such as slavery. In the paragraph after the passage quoted above, Feinberg says "One person wrongs another when his indefensible (unjustifiable and inexcusable conduct) violates the other's right..." (Feinberg 1984: 34). Granted that we accept that no one has a right to own slaves, then aiding runaway slaves or otherwise trying to undermine the institution of slavery is not to wrong slave owners: it is not to violate their rights because they do not have a right to own slaves. An interest in slave-owning is indefensible. This suggests:

  HP: Do not harm others by invading, and so setting back, their rightful interests.

It is thus only the 'rightful' interests of others that are to be protected by morality; it is only harms to these interests that are also *wrongs* which morality forbids.

Setting back an interest that someone does not rightfully possess does not, according to HP, thereby wrong the person.[4] Setting back an interest that the person *does* rightfully possess is wrong but it may turn out to be *justified*. For instance, causing a little pain to a person, and so setting back her interest in not being in pain, by giving her a flu shot may prevent much more pain, associated with getting the flu: assuming that the subject consented and knew what was going on, the prevention of future pain counts as justification. The original judgement of wrongdoing will be withdrawn—I will have more to say about justification later. To conclude this section, we need to ask what it is that distinguishes those interests rightly held from those like slave-owning which are not. We have agreed that an activity like

---

[4]It would be possible to adopt a moral system in which rights played a much more prominent role than they do here. However, as I have explained elsewhere, I believe talking about people's right not to be harmed, for instance, would not contribute to our having a better understanding of the issues, and I believe that talk of rights is best explicated in terms of a moral system such as Gert's, see Forge (2012: 115–118) and see my comments at the end of this section.

slave owning is harmful, but it is *never* justifiable, under any circumstances.[5] This suggests the following criterion for inclusion in the class of interests that *are* rightfully held: either they are not harmful at all, or if they are harmful, then the harm they cause can be justified and hence not classed as wrongdoing.[6] Rape, torture of sentient creatures, murder and racial and sexual discrimination are harmful, they are never justifiable and they are always wrong.[7]

1.2 Other Basic Moral Principles?

The content of HP, what it means and what its significance is, has been sufficiently clarified, for the moment at least. We will need to ask about its status and application: is HP binding on all moral agents in every circumstance, or can there be exceptions? But first we should consider whether it is the only basic principle of morality. That is, does morality just enjoin us not to harm others or does it demand more? Morality surely could not demand less: if morality amounts to anything at all, it must be about the prevention of harm. I have assumed that this is the case, but I will say more in the next section. For the moment, let us ask if morality does, or should, demand more. One obvious way in which morality could demand more is:

> PP: Prevent harm to others.

Harm is understood here, as before, in the sense of set-back to interests (PP could be expressed in such a way that this is spelled out and made explicit). One can argue for including PP in our moral system in the following way: the converse of harming is preventing harm, and if harming is morally wrong, is not preventing harm morally right?

HP is a prohibition or proscription: certain actions are proscribed or forbidden, agents are required to forego certain actions. This accords with the comparison made between morality and laws sanctioned by the state. PP, on the other hand,

---

[5]The rules and strictures against harming that are warranted by HP are not be understood as absolute prohibitions—this is implied by the possibility of justification and the withdrawal of charges of wrongdoing. However, interests that belong to the category of those that are not rightfully held generate a class of rules that are absolute: it is never justifiable to own slaves, to rape, to torture, etc.

[6]We are evidently using a broad sense of "interest". On this reading we can say that a dentist has an interest in keeping his patient's teeth healthy. This will certainly be a financial interest—it is her job—and one hopes she also cares that her patients teeth are healthy and so has an interest in this sense also.

[7]These choices can be disputed; for example, some, unfortunately, believe that torture can be justified. Such questions would require separate discussion and would need to be decided on a case by case basis. One would need to look in detail at what counts as an acceptable justification. Animal cruelty, on the other hand, can surely never be justified: it is never wrong, for instance, to invade the interest of someone who makes money from and pleasure in dog fighting.

prescribes certain actions, namely those that aid others by preventing (or reducing or diminishing) harms that might otherwise befall them, and we may note that the law does not require people of aid others, only refrain from harming them. Notice also that the focus of HP is (or appears to be) the actions of the agent, while that of PP seems to be on the *consequences* of actions. PP requires the agent to act in such a way that the outcome or consequences of the act are harms prevented (or reduced or avoided). As such, PP is a particular kind of consequentialist principle and as such is in the tradition of moral philosophy known as consequentialism, as opposed to HP which we interpret here as non-consequentialist.

PP is not the only possible consequentialist principle that has been entertained by moral philosophers. The consequentialist movement, begun in the nineteenth century, by Mill among others, required moral agents to act in such a way as to convey positive benefit on others. In keeping with the formulations used here, we can express such a consequentialist principle as follows

> CP: Act in such a way as to promote the interests of others (that they have a right to), to the maximum degree possible.[8]

CP is consistent with both HP and PP. Promoting the interests of others is compatible with both refraining from harming others and preventing harms. If the focus of attention was always on the relationship between two individuals, moral agent and moral subject A and B, then promoting B's interests would seem to be inconsistent with harming her interests. As we shall see in a moment, this is not the case. However, the others referred to in CP are 'many others', all of those whose A's actions can affect. So promoting the interests of a larger number of people is compatible with harming the interests of a smaller number, and this may be the only way in which to maximise the benefits. It is possible, on the other hand, to formulate a principle which requires the promotion of the interests of others but not at the expense of setting back any others; this would be a conditionally maximising principle.

There are therefore several different kinds of moral system that could be chosen, and there continues to be discussion and debate in the literature over these issues. A non-consequentialist system based on HP and fleshed out in terms of the sorts of rules that Gert proposes, looks to be the most modest and least demanding on the agent. That it is possible to meet the demands of morality is surely important, because if morality is too demanding then this will be a big disincentive, and hence there are pragmatic grounds for just sticking with HP. On the other hand, a moral system must be demanding enough to make a difference: if morality demands very little, then it may be easy for agents to abide by it but life will not be much better.

---

[8]The classical formulation of Mill runs as follows: "The creed which accepts as the foundation of morals, Utility, or the Greatest Happiness Principle, holds that actions are right in proportion as they tend to promote happiness, wrong as they tend to produce the reverse of happiness. By happiness is intended pleasure and the absence of pain; by unhappiness pain and the privation of pleasure" (Mill 1954b: 6). See Gert (2005: 124–126) for his argument against the promotion of pleasure being a general moral rule.

We do not have to choose a moral system here that will accomplish all the goals and demands that one might wish for. The aim of the present inquiry is to show that weapons research is immoral and unjustifiable, and hence that no moral person should engage in weapons research. I believe that *all* of the kinds of moral systems entertained here, be it based on HP, PP or CP, will deliver the judgement that weapons research is immoral. Here I want to affirm that what I will call common morality has not be chosen because it is the only moral system that will deliver the conclusion that I am after.

1.3   Common Morality and the Harm Principle

Common morality will be understood as a non-consequentialist system based on the HP, our version of the harm principle. Specific rules against harming, such as Rules 1 and 2 of Gert's system, will be sanctioned by HP, where the rules specify a particular kind of set back to an interest. As we have seen, being in pain is clearly harmful to the subject and her interest in being pain-free, and all that entails, is set back. It is an option to identify the system with a set of such rules, as Gert does, but rather than go to the effort of making all such rules explicit, and thereby robbing the system of some flexibility for adaptation to future contingencies, I think it is preferable to take Gert's rules against harming as important elements of the system but which are not all there is to it. The suggestion remains, however, that if harming is morally wrong, then preventing harm is morally right, and since there seems to be a kind of compelling symmetry here, should this positive demand not also be part of the system? We should follow Gert in rejecting such a proposal.

Refraining from harming others can be achieved simply by not engaging in acts that cause harm. It may sometimes be difficult to know if what one does will lead to harm in the future, but if agents are reasonably conscientious in trying to assess the outcomes of what they do, then on the whole they will be able to abide by HP. But to prevent harm, agents must seek out occasions where harms occur, or in danger of occurring, and try to forestall whatever it is that is causing the harm. Where should one begin? There are many places in the world where harm is endemic, where there are wars, food shortages, health crisis, poverty, and so forth. If one is *obliged* to prevent harm, then this will be a full-time job. If the prevention of harm is *demanded* of moral agents, then it seems that unless they devote their lives selflessly to working for others, people will not live up to their moral obligations. Even if an attempt was made to live up to PP, this would necessarily entail being selective and this raises a problem with *impartiality*.

Gert tells us that impartiality is a relational concept: one is partial or impartial in regard to a particular group and in a given respect. And he reminds us that moral impartiality should not be such as to restrict the membership of the group to family, friends, members of the same nationality, race or religion: no one should be excluded from the group of persons protected by morality (Gert 2005: 131–155). So if acting morally consists in preventing harm to others, then there is no restriction to be placed on the group of others. We have just seen what a daunting task it would be if one has to abide by PP. If we are *required* to be impartial when it comes to morality, then PP cannot be the basis of our moral system since it is impossible to abide by it, and the same is the case for CP. If this is correct, then it gives us grounds for restricting morality to proscribing certain forms of conduct, because the notion of prescribing conduct with respect to the group of all moral subjects seems to ask for the impossible. This is not to say that there have been no attempts to deal with this problem by consequentialists and moderate the positive obligations to be imposed on moral agents, but this is not the place to review such attempts.

Harming others is morally wrong, according to common morality, and we have seen that causing pain, disabling, depriving of freedom, not to mention killing, are all varieties of harm. Let us say that the act of breaking the rule warrants the judgment that wrongdoing has been committed, and call this a *prima facie* judgement because it is made before any further examination of the circumstances surrounding the supposed offence has been made, it is an initial or 'first made' assessment. If we interpret the rules of common morality to be absolute prohibitions against harming with no exceptions allowed under any circumstances, then it would seem that the judgment must stand regardless. It must stand regardless if it is correct in the sense that the agent did in fact contravene the rule and if she did so knowingly and voluntarily. If the agent did not do the deed, and the judgement was mistaken in that respect, then clearly it must be withdrawn when this is discovered because the agent has an *excuse*. Coercion is also supposed to excuse, so if the agent is forced the break a rule, then this would count as an excuse as well, and there may be other things that serve as excuses. However, if moral rules are taken to be absolute prohibitions on harming, there can be no *justification* for breaking a rule: once it is established that the agent is responsible for rule breaking, the judgement of moral wrong-doing cannot be withdrawn. It is not a good idea to interpret morality in this way.

There are many examples in which, for instance, some pain must be inflicted to save the subject from much more pain in the future. Going to the dentist is a favourite example: when the dentist drills a tender tooth to patch it up and prevent decay and eventual loss, we don't accuse her of moral wrongdoing. For any moral rule, it is not impossible to think up examples in which breaking the rule is justified. Even the first of Gert's moral rules, the prohibition of killing, could be broken under extreme circumstances, when a person is in terrible pain from a terminal illness. The agents in these examples, dentists and doctors for instance, do not have an excuse for causing pain, etc., because they do so intentionally and are hence responsible. But they have a justification, which is that they are preventing greater pain, suffering, etc. What counts as justification is thus not something 'external to' the rules: the agent is justified in breaking a rule precisely because she is preventing the agent suffer more harm. We can put the point more generally by referring back to HP. HP mentions the rightful interests of a person. Harm is caused when one of more of these interests is set back, but such intervention in the interests of another is justified if her interests (as a whole) would be further set back without the intervention. It is perhaps more perspicuous simply to say that the only justification for causing harm is to prevent more harm down the track. This leads to a justificatory principle:

> JP: The only justification for causing harm by violating the interests of moral subjects is the prevention of further harms.

A *prima facie* judgement that an agent has committed wrong-doing when she has broken a moral rule will be withdrawn if the agent has an excuse or a justification in accordance with JP. In the latter case I will say that the act was *morally permissible*. There is another possibility which we should consider, and this arises when an agent prevents harm without causing any additional harm. For instance, suppose a swimmer rescues another from difficulty in dangerous surf; she is not a lifeguard, so it is not her job to do so, she just helps another person. Had PP been chosen as an element of the framework of common morality, then this act would have described as morally right or morally praiseworthy. But PP is not part of the system, and so there is no warrant for such a designation. There is no reason not to praise the rescuer or say that she is a good person and deserves our praise, but we cannot say she did the morally right thing or that her action was morally good. Gert's system of common morality includes what he refers to as *moral ideals*, as well as moral rules, which urge the prevention of harm. Moral ideals, unlike moral rules, are not binding on the agent in the sense that if she fails to prevent harm, she has not done something morally wrong—his discussion here is similar to our commentary on PP. And preventing harm does not, for Gert, imply that the action was morally good.[9] In common morality, as it is understood here, acts may be classified into three types, those which harm without justification and are morally wrong, those which harm with justification and are morally permissible, and all other acts which are morally indifferent (but which nevertheless could be praiseworthy, estimable, courageous, etc., if they prevent harm).

To finish this account of common morality, let us ask why it is that agents should be moral: what is in it for them? First of all, there is no punishment for moral wrongdoing comparable to the penalties that the law exacts, and in that sense, being a moral person is voluntary. Some moral wrongdoing, such as killing and physically assaulting others are criminal offences and well as moral ones, but some are not. When it comes to 'minor' wrongdoing not punishable by law, such as lying, cheating and being cruel to others, we have a negative attitude towards the perpetrators; we don't think they are nice people and we tend to avoid their company, (as we do for those that commit major wrongdoing). To say this may just be express a kind of folk opinion, and there may well be groups in society where taking advantage of others wherever one can is acceptable. However, there is a view about

---

[9]Gert does allow some actions to be judged morally right—for instance when an agent resists a strong temptation to break a moral rule, Gert (2005: 326). But this is a minor point.

moral wrongdoing that has some currency in the literature that is called "reactive attitudes", which accords with the remark made above, that we react poorly to people who break moral rules and we are justified in having negative attitudes towards them (Forge 2008: 92–93). One reason not to commit moral wrongdoing (and getting found out) is that this puts us in bad odour with others.

Another reason is that if everyone refrained from harming everyone else, then everyone would be better off, because no one wants to be harmed. At the beginning of this discussion of morality, we referred to Mill and the Harm Principle which he thought was the only basis on which the state should interfere with the freedom of action of the citizen. If this is correct, then one assumes that citizens will assent, otherwise they will dispute the power of the state. The state requires that its citizens follow its dictates; otherwise it forfeits sovereignty. Common morality is based upon HP, our version of the harm principle. It covers some of the same ground as Mill's principle (we suppose) but is more inclusive—harms that would not attract penalties under the criminal law, like lying, are forbidden, and moreover some national laws may be outside the scope of common morality. So if people are willing to assent to state regulation of their behaviour for their own well-being, why should they not further limit their behaviour in accordance with a moral system that is practical and not too demanding?

1.4 Conclusion

In the second section of this chapter I said that weapons research can be seen to be morally wrong and morally unjustifiable on the basis of different accounts of morality, accounts that give conflicting judgements about other issues and other matters. It is easiest to make the case against weapons research on the basis of HP and common morality, which is why I use the system here. What I take to be more demanding systems of morality, such as the various forms of consequentialism, I claim give the same judgements about weapons research because they too include (versions of) HP. One can say that the judgement that weapons research is immoral is a product of the common ground between these systems, of shared principles. So I believe we now have in place an appropriate moral framework in which to address weapons research, one that has not been especially tailored so that the desired result will be obtained.

References

Feinberg, J. 1984. *Harm to Others*. Oxford: Oxford University Press.
Forge, J. 2008. *The Responsible Scientist*. Pittsburgh: Pittsburgh University Press.
Forge, J. 2012. *Designed to Kill: The Case Against Weapons Research*. Dordrecht: Springer.
Forge, J. 2018. *The Morality of Weapons Design and Development*. IGI: Hershey, PA.

Gert, B. 2004. *Common Morality*. Oxford: Oxford University Press.

Gert, B. 2005. *Morality: Its Nature and Justification*. Revised Edition. Oxford: Oxford University Press.

Mill, J.S. 1954a. *On Liberty*. London: Dent.

Mill, J.S. 1954b. *Utilitarianism*. London: Dent.

Chapter 2
The Nature of Weapons Research

In order the make the case against weapons research we need to know what weapons research is, and the aim of this chapter is to explain what it is. There is no need to come up with an absolutely precise definition which includes every instance of weapons research and excludes everything else, a kind of 'demarcation criterion'. We do, however, have a choice as to how we understand "research", and this will become clear after I have introduced two examples. One of these, the Manhattan Project, is familiar to many, at least in outline, and has been the subject of much writing and discussion. The other, the development of the torsion catapult in the fourth century BCE, is much less familiar. I have chosen these examples for several reasons. The first is that I want to suggest that weapons research is not a new or recent phenomenon, but has a very long pedigree—this suggestion is a consequence of how I think we should understand research. The Manhattan Project was the beginning of the nuclear age which led to the nuclear-armed world we live in today, with enough nuclear weapons to extinguish much of sentient life on the planet. No more need be said about its relevance to the present discussion.

There are many more recent weapons research programmes which could be used as examples, but it is preferable to chose those that have, so to speak, a history, and where their impacts can be seen and evaluated. This is true, of course, for the Manhattan Project which produced weapons that killed many hundreds of thousands of innocent civilians and ushered in the nuclear age. It is also true in the case of the torsion catapult. This innovation was taken up by Philip and Alexander of Macedon and helped them achieve hegemony over the Greek city states. It was also eventually taken up by the Romans, who incorporated catapults into their legions, primarily as siege artillery, and it was also used as naval artillery.

Weapons research takes place in a particular *context*: weapons research requires skilled researchers, workers, and it requires resources, facilities and materials. The location of a weapons research programme will therefore be a particular place (or places) and times where materials and workers are available. I will however understand the *context* of weapons research to have other elements which will, for instance, provide a basis for explaining why the programme was set up, and as such

© The Author(s), under exclusive licence to Springer Nature Switzerland AG 2019
J. Forge, *The Morality of Weapons Research*, SpringerBriefs in Ethics,
https://doi.org/10.1007/978-3-030-16860-5_2

we would expect these extra elements to have to do with the political and historical circumstances of the time. As we shall see, the context in which weapons research is undertaken and the contexts in which the weapons are used may be, and often are, different.

2.1 Manhattan Project

A great deal has been written about the Manhattan project and I have nothing new to say about it.[1] It is the most famous, or notorious, weapons research programme and it is especially relevant for present purposes in regard to its genesis and its aftermath. The Manhattan project was set up to see whether it is possible to make a weapon out of the heavy metals uranium and plutonium. There had been specu- lation among some physicists in the 1930s about whether recent developments in nuclear physics, such as the discovery of nuclear fission, might have military applications, for the following reason: when the nucleus of a heavy element breaks apart, or fissions, it releases a small amount of energy as well as a number of neutrons—neutrons and protons are the constituents of the nucleus, having (about) the same mass, with the latter positively charged and the former bearing no charge. Fission was observed on several occasions in the 1930s, but it was not identified as such until 1938 in Berlin by the German physicists Otto Hahn and Fritz Strassman. Since fission is caused by neutrons striking a nucleus and since fission liberates neutrons, this raises the possibility of a *chain reaction*: one fission causing another fission, and so on. And even though the amount of energy liberated in each such event is very small, if enough fissions take place, this could add up substantially.

This idea of a nuclear weapon was what the Manhattan project set out to investigate, but it had been thought up about ten years earlier by Leo Szilard— indeed, the *idea* of an atomic bomb is usually credited to Szilard (Forge 2012: 83– 87). Szilard was Jewish and left Germany in 1933, around the time a number of other German scientists departed, just as the Nazis were coming to power. It was primarily these scientists who not only speculated about what might be possible on the basis of nuclear physics but tried to set up nuclear weapons research projects to test their theories and to draw these matters to the attention of governments, first in Britain and later in the United States. The British government became involved first and considerable progress was made in Britain up until 1941, when it became clear that enough of the fissile material, either of the requisite isotope of uranium or plutonium, which must be made in a nuclear reactor, could not be obtained under

---

[1]I still think Rhodes (1986) is the single best and most readable source. Hoddeson et al. (1993) is a non-technical discussion of the period between 1943 and 1945, when the designs for the bombs were developed, and is based on original documents. Smyth (1989) is the official report of the US government's involvement from 1940–1945, a blow by blow report in numbered paragraphs. Finally Serber (1992) is a set of lectures on how to build a bomb—in outline and with no values for the design parameters—written by a physicist who was at Los Alamos.

wartime conditions. However, two émigré German physicists working at Birmingham University, Otto Frisch and Rudolf Peierls, provided convincing theoretical arguments that a chain reaction could be sustained for long enough for a large amount of energy to be released. This information, and more besides, was given to the Americans in October 1941. Shortly thereafter, President Roosevelt authorised a programme to see if an atomic bomb was possible.[2]

The Manhattan Project was formally set up in August 1942. And work got underway to build an nuclear reactor to produce plutonium, a plant to separate the preferred isotope of uranium and finally a weapons laboratory at Los Alamos New Mexico. So the quest for an atomic bomb moved from being essentially a university research programme, which it had been from 1940, to an industrial project. What sustained the research in the US until the end of 1941 were the events in Europe, such as the German invasion of the Soviet Union in June of that year. Even if the US remained neutral throughout the war, it was still prudent to look into the possibility of a new weapon which the Germans might acquire. The main focus of much of the university research in 1941 was the rate at which neutrons were generated in an assembly of fissile material, as this was the defining parameter for a chain reaction. I will explain briefly.

An assembly of fissile material will have a given mass, volume, shape and constitution in terms of the proportion of isotopes of the element in question. In the case of uranium, this will be a mixture of uranium-235 and uranium-238, the former being the more fissile and the latter being (by far) the more naturally abundant. Samples of (artificially) enriched uranium have more uranium-235 than occurs in nature. Since such assemblies are made up of fissile material, atoms will (by definition) fission when bombarded with neutrons. Suppose this happens. Then after a given time, there will either be more neutrons in the system, less, or the same number, and these conditions correspond to a divergent chain reaction, a convergent chain reaction, and a steady state. Enrico Fermi, who oversaw much of this research, represented the situation as follows: suppose k is the number of first-generation (introduced) neutrons, k^2 the number of second-generation neutrons caused by fission, k^3 the third generation, etc., then if $k^n/k^{n-1} > 1$ for sufficiently large k there is a divergent chain reaction, $k^n/k^{n-1} < 1$ the reaction converges, etc., for the first generation fission events. A divergent chain reaction is what is necessary for an atomic bomb. It does not follow that if a divergent reaction is observed, that it will continue to diverge, diverge quickly enough or for long enough, but if no divergent reaction can be achieved under any conditions for any fissile material, then a bomb will not be possible. Fermi's team observed $k^n/k^{n-1} > 1$ in December 1941, in what was in effect the world's first nuclear reactor. At a university on the other side of the country, in Berkeley, research was being done on the tiny quantities available of the

---

[2]Rhodes tells us that there is no formal record of the decision to go ahead and actually try to build a bomb, but undoubtedly the Japanese attack on Pearl Harbour in December 1941 would have played an important part.

other fissile material, plutonium. The results, such as they were, suggested that it was even more fissile than uranium-235 and hence a more potent bomb material.

Given that an atomic bomb had not been ruled out because the properties of the candidate materials were observed to be unsuitable, the overall agenda for the Manhattan Project was the following. Enough uranium-235 and plutonium needed to be made available for further testing, and if that proved promising, for manufacturing atomic bombs. To that end, large facilities were built in the states of Tennessee and Washington. Further research needed to be conducted on the basic properties of uranium-235 and plutonium to see whether assemblies could be created in which enough generations of neutrons could be produced quickly enough to create an explosion of unprecedented size and destructive power. Finally, if that were possible, a weapon had to be designed and made. As it turned out, three weapons were made on the basis of two different designs by July 1945. One was tested and two dropped on Japanese cities. The designs in question could be expressed or formulated in a variety of ways—we have an idea about what these would be like, but no complete design is publicly available, for obvious reasons. The designs are highly technical, with statements of the values of crucial parameters and relations between them expressed in mathematical terms. Any group with the requisite skill and access to resources, including fissile material, would be able to manufacture an atomic weapons on the basis of these specifications. That is the very nature of a design: it is not something that can only be used once at a given time and place. These is no limit to the number of artefacts that can be produced on the basis of a given design, nor is there a limit on when and where they can be made.

2.2 Catapults

One would not necessarily expect an Oxford professor of classics to build a catapult, but E. W. Marsden did (Marsden 1969: 86).[3] This was not for fun but to test the instructions set down in a treatise on Greek artillery published in the first century CE and written by Heron of Alexandria. Heron wrote two such 'technical treatises', and his were the last of a series of five, the first of which was written between 270 and 240 BCE. Marsden translated all of these, from Greek and Latin (Marsden 1971). He also wrote a pioneering work on Greek and Roman artillery (Marsden 1969). If Marsden was able to build a catapult by following Heron's instructions, then this is sufficient for concluding that Heron's treatise set down a design for the machine. The aim of a design is to enable the artefact to be produced, more or less exactly as intended, and hence if the artefact is produced, then this is enough to conclude that a design was available. Given that Heron's instructions

---

[3]I referred to E.W. Marsden as A.E. Marsden in Forge (2012). Here I correct that mistake. Also, as in Forge (2012), I am using the word "catapult" to refer to a range of different machines for casting both stones and arrows (Forge 2012: 39).

were sufficient to build a catapult, it is no surprise that one could be built two millennia later—we do, after all, have much greater knowledge and expertise—although Marsden reports he had to use rubber instead of hair in one part of the machine and he thought that this reduced its performance. All of the treatises were designs for making torsion artillery. I will briefly explain what this is.

All forms of artillery are means for throwing, casting, shooting or firing projectiles. An essential difference between artillery and other ways to cast or fire projectiles, such as by means of slings or bows, is that the energy required to accomplish this can be stored. Since the discovery of gunpowder, the energy has been stored in bags of powder, bullets, shells, etc., and artillery functions as a mechanism for aiming the projectile and releasing stored energy. Ancient artillery stored energy either as the tension in a string, as was case with tension artillery, or as the torsion in a mass of 'springs' which were either rope or hair. The term "torsion" signifies that something is twisted. Think of a thick length of rubber that is twisted: the more turns, the harder the effort needed, and once released, the rubber will untwist itself to the original shape. In the twisted state, the work needed to impart the twist is stored in the material. Moving to torsion artillery, the basic idea can be grasped in the following way. A mass of ropes is wound round a stout piece of wood, which I will call the base, and another piece of wood is buried in the ropes in the middle, like a handle. Moving the handle will simply rotate the base. Now suppose the base is firmly held in a frame in the horizontal direction If the handle is set perpendicular to the base, pulling down on the handle will require some effort to overcome the resistance of the ropes. On release, the handle will then spring back to the vertical. This is the operating mechanism of the simplest form (one-armed) of torsion artillery, known as the mangonel. The machine becomes an artillery piece when elements are added which allow for stones or arrows to be fired when the handle is released (Marsden 1969: 18–19; Rihill 2007: 77).

A set of instructions that incorporated an expanded (and clearer) description of the torsion mechanism just outlined, together with details about the dimensions of the structure, would enable competent artificers to build a catapult. Indeed, the technology spread from the Greek city states from the fourth century BCE right round the Mediterranean, until torsion artillery was found in all significant military formations, and by the first century CE was incorporated into the Roman legion and mounted on warships, such as Caesar used in 55 BCE when he invaded Britain (Wintjes 2016: 21). However, the machines were not uniform: a wide variety of shapes and sizes were found, from larger siege artillery, to smaller pieces suitable for artillery towers in fortified towns and for armies in the field. To some degree, this was no doubt due to the skill and flexibility of the artificers, but it was also due to the fact that not only did the Greeks invent torsion artillery, they identified a key parameter that enabled the catapults to be scaled up or down, and this was the diameter of the base. On reflection, we might conclude that this is not that surprising. After all, the wider the base, the more rope could be used, which in turn could support a more massive handle that could be used to launch a bigger projectile. However, the Greek weapons designers discovered two formulae which optimised their machines for given missions. For example, suppose effective siege artillery needed to be able to throw stones of a certain weight at city walls in order to effective. If the missiles' weight should be W or thereabouts, then the following formula specifies the optimal diameter D of the base

$$D = 1.1^3 \sqrt{(100\,W)}$$

here $^3\sqrt{}$ signifies a cube root. This amounts of an instruction to build the right-sized catapults: build a frame that can securely house a base of diameter D and add the additional elements needed accordingly.

It is evidently a great advantage to be able to construct artillery best suited to the mission for which it is intended, without wasting time and materials experimenting and improvising. The formulae were therefore no doubt extremely useful. And they are also quite remarkable. The Greeks did not know any theories about torsion or torque, nor did they know any theory about the winches needed to pull back the catapult handle. There is, in any case, no theory that allows for the deduction of the base formula. Not only this, the Greeks had no algebraic method of solving the formula for given values of W—for finding the cube roots—which had to be done using geometrical approximation (Marsden 1969: 39–41). One person who we know was able to do this was Archimedes, who constructed a wide range of catapults for the defending the city of Syracuse from the Romans, and no doubt other engineers were able to do so as well (Steele and Dorland 2005: 1–7). The base formulae must have been arrived at by careful experiment. The only reason why we would *not* classify this work as weapons research is if we had stipulated a priori that weapons research must be explicitly informed by scientific theory. In that case there would have been no weapons research before the nineteenth century. This is arbitrary and it is clear that any historically-sensitive inquiry into the subject will count the development and evolution of the torsion catapult as weapons research.

2.3   Weapons Research and Weapons Design

On the basis of the example just discussed, we can say weapons research has been in existence for least 24 centuries. There is, as I have just suggested, no good reason why weapons research must be tied to scientific practice as it is understood today. And even if the Greeks had known about torsion, torque, angular momentum and related concepts, and so understood how torsion catapults worked, understood the theory, it would not have made a great deal of difference. The careful experimental work would still have been necessary in order to find a relation between D and W that worked in practical terms. Granted that what is important are outcomes of weapons research, artefacts and designs, then the former could be characterised as whatever achieves its aim: if by some process a design for a weapon is forthcoming, then that is, by definition, weapons research. If we adopted this way of looking at weapons research, then there is evidence that the timeframe of the activity should be extended much further back.

By far the most sophisticated, and expensive, weapons system before the advent of artillery was the chariot: and it first appeared in Central Asia at the beginning of the second millennium BCE (Lee 2016: 59). Chariots were drawn by two or four horses and were normally manned by a driver and a warrior equipped with a bow or sometimes with a lance or sword. Chariots were fast and mobile and they dominated the battlefield for eight centuries and, according to Lee "advantaged those state-based civilisations that could martial the necessary resources and labour to make large forces of this new vehicle" (Lee 2016: 81). The Hittites of modern Turkey belonged to one such state, which was able to defeat the Egyptians in the Battle of Kadesh, in which it is estimated that there were at least seven thousand chariots. Chariots spread from Central Asia to the Near East, to China, India and even to Europe, but there was nothing comparable to the technical treatises for the catapult to formalise and codify their designs. Nevertheless, chariot technology did indeed spread; the technology was modified for local conditions, but the spoked wheel design, for instance, was ubiquitous. Chariots were made by skilled craftsmen and these people would have been the keepers of the knowledge of how to build them. The chariot evolved from, and was designed on the basis of, the war cart.[4] The latter were depicted on artefacts from Sumeria dating from 2440 BCE. These were not very formidable weapons because they were slow and heavy, and their main function is thought to have been to bring kings, aristocrats and other important persons to the battle, but the chariot was formidable.

I believe the evidence in favour of the chariot being the product of conscious and systematic attempts to develop a fast mobile weapon using horse power on the basis of an analysis of the limitations of the war cart, namely being an example of weapons research, is compelling. And there grounds for expanding the time frame of weapons research still further. Sophisticated compound bows made of wood, sinew and horn have been around for many thousands of years, as have weapons made of iron and bronze. There is, once again, nothing comparable by way of evidence of design to the technical treatises for torsion artillery, save the weapons themselves. These constitute evidence of design because they were, if not identical, sufficiently similar to suggest that they were all made according to the same method and to the same standard, and that is the hallmark of design—unique, one-off creations by an artisan or artist are, for that reason, not a basis for the reproduction of many copies and hence are the antithesis of design. Again, I believe that there is sufficient evidence to argue that weapons research has been around for a very long time, perhaps at least ten thousand years, in one form of another.

It may, however, be asked whether it really matters just how old weapons research is. If our overall aim is to argue that it is morally wrong, does it really matter how long it has been going on for? I think that it does matter. As I have said

---

[4]To refer again to Wayne Lee "Soon thereafter, not later than about 2100 BCE, some of them [dwellers just southeast of the Urals], perhaps having seen a Mesopotamian war cart, grasped the advantages of being elevated about the battlefield, standing to fight, and using their horses to draw the vehicle quickly while using only two wheels, a design that allowed the vehicle to turn much more quickly. In short, they developed a true chariot" (Lee 2016: 57).

already, the fact that an activity has been routinely undertaken for a long time does not mean that it cannot be wrong nor that this cannot be realised and the activity proscribed. It matters that philosophers try to engage with such routinised practices. Moreover, as we shall see, and have seen already in this chapter, the extended timeframe of weapons research provides clear examples of how the products of weapons research conducted in one context find application in other contexts. And there is a bigger picture as well. Weapons research has aided expansion, conquest, colonisation and empire-building for at least the last two millennia and a half, and in so doing has abetted crimes against many innocent indigenous peoples, and this should be kept in mind.

Even though weapons research has a long history, it did not produced truly dangerous weapons until nineteenth century. In battles and wars up the American Civil War, the majority of military and civilian deaths in war were due to disease and lack of food. In the middle of the nineteenth century, rifles that were accurate to 500 m became available and vastly increased battlefield deaths.[5] It has not been apparent until quite recently just how lethal weapons can be, and it has not been apparent how the products weapons research have transformed warfare from relatively short-lived, infrequent and isolated engagements to large drawn-out conflicts. For example, the major naval engagements in the whole of the Napoleonic Wars lasted less than a week all together, while the siege of Leningrad in the Second World War lasted over two and half years. However, a global thermonuclear war with weapons spawned by the Manhattan Project would be both lethal, in respects it is hard to imagine, and of very short duration.

Returning to the idea that when a process leads to a design for a weapon, that is enough for it to qualify as weapons research, there are two reasons why this should not be adopted as the *definition* of weapons research. In the first place, there can be unsuccessful weapons research. German scientists in the Second World War explored the *possibility* of an atomic bomb and concluded that it could not be made to work—a crucial mistake was made in the estimation of the neutron multiplication factor k. To say that the scientists were therefore not engaged in weapons research seems quite wrong. The second reason is that research that makes a contribution to weapons research but which does not aim to do so would therefore be counted as such and this would very greatly expand the range of the activity to the degree that it may be hard to see what is not weapons research. As an example, nuclear fission was discovered by Hahn and Strassman in Berlin in 1938, and this was an important step along the road to the atomic bomb. But Hahn and Strassman were not working in a weapons laboratory, their work was not sponsored by the military, it was

---

[5]A soldier shot and killed by a musket ball in battle was very unlucky. But when it came to the American Civil war, the estimates are that of the 200,000 solders killed on the battlefield and 450,000 wounded, 90% were caused by rifle bullets, 9% to artillery and 1% to bayonets, swords and other arms (Howey 1999: 50). Many of those wounded died from their wounds and disease. Nevertheless, the idea that Civil War commanders and their soldiers immediately understood how to fight with these new weapons and so abandoned the age old tactic of closing up to the enemy to exchange volley fire, has been discredited (Lee 2016: 368).

openly published in scientific journals, and they had no intention to provide information for a bomb project which, by all accounts, they had no knowledge of.

What is lacking here is reference to the *intention* of the participants.[6] Weapons research is intentional action: participants know what they want to achieve and act in ways that they believe will achieve their aims. This determines how the actors represent to themselves what they do. Weapons research is an activity that actively seeks designs for weapons: this is what it aims to do and this is how it is to be defined. I will argue that weapons research is immoral and unjustified, and this implies that moral agents should not undertake weapons research. That is to say, whatever moral agents do, they should not intend to do weapons research, they should not conduct research that aims to produce designs for weapons. The moral imperative that I want to establish thus engages directly with the desires and goal-directed behaviour of rational persons. My definition of weapons research is therefore as follows:

> Weapons research is research carried out with the intention of designing new weapons or improving the design of existing weapons or designing or improving the means for carrying out activities associated with the use of weapons.[7]

I will have more to say about designer intention in the next chapter.

2.4   The Context of Weapons Research

Weapons research is applied research. This is to say that weapons research is not an 'end in itself', inspired by curiosity and done out of pure interest. The latter is called pure or basic research, and clearly weapons research is not of this kind because it aims for some specific end external to the activity, namely designs that can be used for making new or improved weapons. In my book on science and responsibility I argued that it is the *context* in which research is conducted that determines its character as applied or basic, not the actual substance or content of the work (Forge 2008: 16–20). There are many examples of work carried about both before and after the setting up of the Manhattan Project that was done with a view to seeing if an atomic bomb was possible and which were therefore instances of applied research. But the self-same work conducted out of pure interest would have been basic research. As it happened, Fermi's work on the neutron multiplication factor k was applied. On the other hand Hahn and Strassman's work on fission was basic, but it was reproduced and confirmed in the context of the bomb project and hence in that instance qualified as applied. The outcomes of weapons research endure long after

---

[6]For an account of intention and its relevance to present concerns, see Forge (2008, Chap. 5).

[7]I first used this definition in 2004, see Forge (2004: 534), and have employed it on a number of occasions since. The activities associated with the use of weapons include their command and control.

the context in which the work was done has been transformed—torsion catapults can be built today on the basis of instructions written two thousand years ago—and this fact about weapons research will be crucial in denying that the practice can (ever) be justified.

For there to be any kind of research, basic or applied, there must be individuals willing to undertake the work and resources available for them to do so. The difference between basic and applied research is that in the latter case the resources can be viewed as investments provided in the expectation of some future (external) gain. By examining how and why resources are devoted to applied research projects, we will be able to better understand why they are carried out. In regard to weapons research, it is a truism that the activity is supported because those with the capacity to do so want new or improved weapons. However, looking at individual episodes of weapons research more closely and asking just why new or improved weapons are desired at that time and place is a different matter. We have seen that the Manhattan Project was set up to see if an atomic bomb was possible just in case Germany was able to build one. The United States was at war and it wanted to make sure its most technically advanced adversary did not have a monopoly on a terrible new weapon. There is no information about exactly when the torsion catapult was invented; the technical treatises are not reports of actual experiments. But it is generally agreed that the two-arm torsion catapult, invented after the simpler one-arm machine described above, was developed under the auspices of King Philip II of Macedon, Alexander the Great's father (Rihill 2007: 80). Philip and his son were aggressive rulers bent on conquest and for that reason wanted the most advanced weapons.[8] Chariots were expensive and only states with sufficient resources were able to acquire them in large numbers and these states became dominant in the second millennium BCE.

Suppose it has been established that weapons research is morally wrong and we then ask if it is justified, and hence if the activity is permissible. It is clear that there can be no justification for the activity as a whole or in general, just as there cannot be for any kind of action deemed morally wrong. Whether justification is possible will depend on the particular instance in question, and it would appear on some, if not all, such occasions, justification will need to appeal to the context in which the research is conducted. The reasons why the programme was set up will be a basis, perhaps the only plausible basis, for constructing a justification. For our two examples, only the first could be justified. Philip wanted the most effective weapons to extend his rule over the whole of Greece and Alexander took over Philip's armies and conquered lands all the way to India. Aggression and conquest will not serve to justify moral wrongdoing, that much is clear. The first example is different. Research into the atomic bomb was intended as a way to find a response to any successful German project, and in the war it was Germany and its allies who were

---

[8]It is worth quoting Rihill "King Philip of Macedon, the usual suspect in this case [invention of the torsion catapult], was born into a family whose history revolved around the ruthless pursuit of power and was littered with battlefield deaths and political assassinations" (Rihill 2007: 80).

the aggressors. Both Britain and the US were attacked and undertook weapons research in order to find ways to defend themselves and fight back, and the Manhattan Project and the earlier work done in Britain were part of this defence effort.

I am going to argue that even though defence against aggression is permissible, not every measure taken to aid the defensive effort is allowed, including weapons research. This may seem implausible and I will of course need to set down my reasons. However, reflection on the context of weapons research will give an indication of what these are. Weapons designs do not live and die in the context in which they were created; unlike the hardware that they enable, designs do not have a limited shelf-life. We have seen that torsion catapults can still be built from the designs set down in the technical treatises and it was mentioned that this innovation spread round the Mediterranean. Indeed, the Romans used torsion siege artillery in their conquest of the Greek cities states in the middle of the second century BCE, so a weapon that was invented in Greece was used to subdue that country. This was clearly a new context. Philip and Alexander had been gone for nearly two centuries and a new power had risen which used their own weapons to conquer their country. What happened to the products of the Manhattan Project is well-known. By August 1944 it was clear that the Germans never had a viable atomic bomb project (see Goudsmit 1947: 3). Before the Manhattan Project had produced any atomic bombs Germany surrendered, and the weapons were used against Japan. Here again the context had changed in the sense that the rationale for investigating the possibility of nuclear weapons no longer applied. The Manhattan Project was the catalyst for the nuclear arms race of the Cold War, but that context also changed with the collapse of the Soviet Union leaving as a legacy vast and terrible nuclear arsenals.

Here I will mention one of the most important contextual factors that conditions the nature and direction of weapons research, and this is *military doctrine*. Here is Barry Posen's characterisation:

> Military doctrines are critical components of national security policy or grand strategy. A grand strategy is...a state's theory about how it can best 'cause' security for itself....I use the term "military doctrine" for the subcomponent of grand strategy that deals explicitly with military means. Two questions are important: What means should be employed?, and How shall they be employed? (Posen 1984: 1–2)

Military doctrine has to do with the means that a state will need to use if its security is threatened to the extent that it has to resort to violence. In other words, military doctrine has to do with the kinds of weapons that a state needs to acquire and hence military doctrine gives rise to weapons manufacture, if the weapons designs are available, or weapons research if they are not. However, military doctrine can change and so ideas about the 'means that should be employed' will also change. Once the Soviet Union collapsed, the United States no longer had to confront another superpower with diametrically opposed interests and ideology. But the means acquired under the doctrines that were 'critical (sub)components' of various grand strategies of both sides, namely the vast array of nuclear weapons, remained. Posen tells us that there are three kinds of military doctrine: offensive, defensive and deterrent doctrines, which pick out different kinds of military operations (Posen 1984).[9] It seems that this covers all the bases: one can use one's military forces to attack, defend or deter, but not much else! I will have more to say about military doctrine, especially in Chap. 4 when I talk about 'levels of strategy', and in Chap. 6 when I revisit the idea of the context of weapons research.

To sum up, introducing new means to kill and destroy into the world is not something to be done without very good reason. This is surely obvious. It is surely something no one could dispute. I have suggested that the very good reasons in question must be something to do with the context in which the weapons research is undertaken, for instance that there is a war on. But contexts change while the output of weapons research endures. As contexts change, the 'very good reasons' may no longer have purchase or application, and hence it is as if weapons research is done *without any* good reason. It is as if weapons research is done without a good reason because the fruits of weapons research remain available in the new context. The development of nuclear weapons is most dramatic example of this, but there are many others, some of which, such as the invention of the Kalashnikov AK-47, I will mention later. The crucial issue for the justification of weapons research is this: do the demands of the present context, the one in which weapons research is set up, outweigh those of the future contexts in which the new designs will still feature? It is question no one can answer!

2.5 Conclusion

The aim of this chapter has been to explain what weapons research is, with my preferred definition given in the third section. While I do not think this definition is controversial, it might be objected that there was no extended discussion about the nature of research and so the wide view of weapon research adopted here was not well-supported. But that objection misses the point. There is no absolutely correct view on research; a little reflection will show there are different kinds of activity that can be characterised as research. We would be at liberty to take weapons research to be a relatively new phenomenon, informed by scientific research, but the only consequence of that decision would have been fewer instances in a shorter time frame, and, I believe a less historically-informed and historically-sensitive

---

[9]It does not follow that states with aggressive military doctrines will acknowledge them, even when they put them into practice. When Hitler annexed Czechoslovakia in 1938 he claimed he did so in response to the persecution of ethnic Germans, who he was obliged to defend. And when Frederick the Great annexed Silesia in 1713, in a bare-faced act of aggression, seeking to raise Prussia to 'Great Power' status, he claimed that he did it in order to make Europe more secure (Kennedy 1987: 110). There are many more examples, some more of which will be discussed in Chap. 4.

discussion. It would not have changed any of the substantive arguments. I have also discussed the context of research, and this is an idea that will need more refinement in the sequel.

Appendix: Weapons Research and Military Technology

I have defined weapons research as research that is concerned with designing weapons and with the means for using them. It might be asked if this description neglects military technology: does not weapons research aim to produce new military technologies? I have no objection whatsoever to this way of describing weapons research; I could have defined weapons research with reference to military technology, but I think this is an oblique way of doing so. It would be necessary to say what technology is and it is simpler, I believe, to refer directly to weapons. Nevertheless, I will do so here, to show that these two ways to thinking about weapons research are essentially equivalent.

The received view of technology is that it is knowledge of technique, where a technique is a way of doing something. Military technology is thus knowledge of techniques that harm living creatures and destroy things that they value. Catapult technology is therefore knowledge of techniques for casting stones and arrows, in order to reduce fortifications and kill their defenders at long range, etc. Military technologies and the weapons themselves are not the same things: the latter are *substantive*, they are material things, made of wood and rope for instance, whereas the former are propositions, items of knowledge. Technology understood in this way—and there are alternatives as we will see in a moment—resembles design in that both are kinds of knowledge, but again the output of weapons research, namely designs, is different from the corresponding military technology. A design for a weapon is typically a set of instructions for making a weapon, the weapon can then be said to 'constitute' the technique, it is what makes the technique practical. Philip and Alexander knew how to besiege fortified places *because* they had catapults and catapults shoot stones and arrows. This is not, however, to say that thinking about technique has no role to play in weapons research itself, far from it. Suppose a weapons researcher in Central Asia at the beginning of the second millennium BCE was thinking of a way to use horses to provide a fast-moving weapons platform— this a made-up story—one that could outrun an infantryman. Our ancient weapons researcher is contemplating a particular technique. On the basis of his experience of war carts, he comes up with the idea of a chariot. The weapons researcher conjures up the technique, he imagines it, and then invents a weapons that makes it possible in practice. I am not claiming that this is always how weapons research is conducted, but no doubt it sometimes is. But, again, the technique is different from both the weapon that enables it, and the design for the weapon.

To define weapons research as an activity that aims to produce new or improved military technologies would focus attention on particular techniques, on casting stones and arrows, on horse-drawn weapons platforms, on massive nuclear explosions, and so on. These ways of harming are executed by means of weapons, and as we have just seen, weapons are the means by which the techniques are put into practice. To define weapons research as aiming at military technologies is thus not to say that it does not aim at weapons design, because to have a military technology in place it is necessary the have the corresponding weapon. This is why I said that this way of looking at weapons research was essentially the same as the one I have chosen. But this would not necessarily be true if some other view of technology were adopted, and one reason why I did not define weapons research with reference to military technology was that I did not want to engage with the various accounts of technology available, except in this appendix, where I will devote one paragraph to talking about one of the more interesting of these, due to Don Ihde.

Ihde's definition of technology in his *Philosophy of Technology* has three necessary conditions (Ihde 1993: 47). Firstly, a technology must have a concrete or material component, and this condition immediately distinguishes his idea from what I have called the received view. In regard to military technologies, one assumes that weapons themselves are the prime candidates for the material part. In addition, there is a set of praxes or uses that people make of the component and a third condition which I will leave aside. So in addition to the weapon in question, a military technology will comprise (all) the uses of the weapon. Thus if a type of catapult is versatile and can be used as siege artillery, field artillery as well as be mounted on ships, then all of these uses, and any others that might occur, would comprise the technology, along with the catapult type itself. What Ihde's definition omits, it seems, is reference to technique. Using the catapult example again, what catapults do from this perspective is cast stones and arrows. It is true, of course, that when a catapult is used as siege, or field or naval artillery, it is used to cast stones or arrows. But the uses are particular applications of the technique, of what the weapon does, and Ihde is explicit about technologies having a set of praxes, not just something they do because of the what they are. But I will leave the matter here. Enough has been said to indicate how the idea, or ideas, of military technology fit with the present approach, which describes weapons research as aiming to design new weapons, and about why this approach is the more straightforward.

References

Forge, J. 2004. The Morality of Weapons Research. *Science and Engineering Ethics* 10: 531–542.
Forge, J. 2008. *The Responsible Scientist*. Pittsburgh: Pittsburgh University Press.
Forge, J. 2012. *Designed to Kill: The Case Against Weapons Research*. Dordrecht: Springer.
Goudsmit, S. 1947. *Alsos*. New York: Schuman.
Hoddeson, L., P. Henriksen, R. Meade, and C. Westfall. 1993. *Crtical Assembly*. Cambridge: Cambridge University Press.
Howey, A. 1999. *Civil War Illustrated Times* 38: 5.
Ihde, D. 1993. *Philosophy of Technology*. New York: Paragon House.
Kennedy, P. 1987. *The Rise and Fall of the Great Powers*. New York: Vintage Books.

Lee, W. 2016. *Waging War*. Oxford: Oxford University Press.

Marsden, E. 1969. *Greek and Roman Artillery: Historical Development*. Oxford: Oxford University Press.

Marsden, E. 1971. *Greek and Roman Artillery: Technical Treatises*. Oxford: Oxford University Press.

Posen, B. 1984. *The Sources of Military Doctrine: France, Britain and Germany Between the Wars*. Ithaca: Cornell University Press.

Rihill, T. 2007. *The Catapult*. Yardley, Penn: Westholme.

Rhodes, R. 1986. *The Making of the Atomic Bomb*. Harmonsworth: Penguin.

Serber, R. 1992. *The Los Alamos Primer*. Los Angeles: University of California Press.

Smyth, H. 1989. *Atomic Energy for Military Purposes*. Stanford: Standford University Press.

Steele, B., and T. Dorland. 2005. *The Heirs of Archimedes*. Cambridge: MIT Press.

Wintjes, J. 2016. "Ancient Naval Artillery Support" in G. Dworok and F. Jacob. *The Means to Kill*. Jefferson, NC: McFarland.

Chapter 3
Weapons Research is Morally Wrong

Weapons are artefacts, which is to say that they are 'man-made', crafted by us for some end or reason.[1] Artefacts are of two basic kinds: the aesthetic and the practical. The former are made to be admired, while the later are made because the help us achieve something, either something that we could not do unaided, or do something more cheaply, more easily, more efficiently, etc. Weapons are practical artefacts: they enable people to harm others more easily, more cheaply, more efficiently, in new ways, on a grander scale, and so forth. I am going to begin this chapter by considering weapons as (practical) artefacts, so we can better understand the kind of things they are. Weapons research, as we have seen, is an endeavour that seeks out new kinds of weapons: weapons that are altogether entirely new, like the atomic bomb and the catapult; weapons that are new variants on a type, such as the German Mauser rifle; weapons that are improvements on an existing weapon, like an improved version of the Soviet M-34 battle tank; or new or improved ancillary structures, such as the chariot and the galleon, first used as a weapons platform in the 16th century Anglo-Dutch wars. Weapons research aims to find workable *designs* for weapons and their ancillary structures.

A weapons research programme is successfully concluded when it comes up with a design which works, when it is possible to build a prototype which does what it should do. Weapons design has this in common with all research that aims to design artefacts, but it is unique in that it is the only activity which aims to produce new or improved means to harm, to kill and injure people and destroy things of value. Weapons design is the only *kind* of research activity about which common morality makes a judgement, and this is that it is morally wrong, as we shall see in this chapter. To begin we need to say something about artefacts and design in

---

[1]A naturally-occurring object, a lump of wood or a rock, something that is not made by us, can be used *as* a weapon, but it was not a weapon before it was used as such. If the rock were secured to the lump of wood and fashioned into a primitive club, then it becomes a weapon before anyone is hurt by it.

© The Author(s), under exclusive licence to Springer Nature Switzerland AG 2019
J. Forge, *The Morality of Weapons Research*, SpringerBriefs in Ethics,
https://doi.org/10.1007/978-3-030-16860-5_3

general, and in particular determine the role of designer intention, mentioned in the definition of weapons research given in the previous chapter, and designer responsibility.

3.1 Artefacts, Purposes and Designer Responsibility

Practical artefacts are made because they enable us to do things that we could not do otherwise. This is not, of course, fortuitous, because artefacts are made to do a certain job. Artefacts are not always successful, in the sense that they do fail to do what they were supposed to do, but in many cases they work and this is why people keep using them. So with this in mind, it seems clear that those who design artefacts do so with a certain job or purpose or function in mind. We would like to be able to do X, so let us make something that will let us do X. Thus what design aims to do is to come up with something that will make X achievable. So it seems that we can say that those who seek to design such an artefact have X in mind: they *intend* to create something that will let us do X. This conclusion is surely so obvious that no one could dispute it. To do so seems to imply that artefacts fulfil the purposes for which they were designed somehow by accident. Although no one accepts that design is always accidental in this sense, there has been a good deal of discussion about whether designer intention tell us all we need to know about artefacts and what they do in the world.[2] Don Ihde has coined the term "designer fallacy" to refer to the claim that the way artefacts are used is always in accordance with what they intended for. He has many examples that show that historically important technologies, like the typewriter and the telephone, have come to be used in quite different ways from those that their designers envisaged. But this claim is entirely compatible with (very) many artefacts being designed for a particular purpose and being used (primarily) for that purpose. That a given artefact comes to have different uses is, as Ihde rightly notes, a consequence of its being emplaced in a given context, and the ways it is taken up in that context may not have been something that the designer could have anticipated.

We can express these insights by drawing up a simple tripartite division of the purposes of artefacts. I begin with an example, the pocket watch. The first or *primary* purpose of a pocket watch is to tell the time; this is what watches do and this is what those who design watches intend them to do, and the more accurately a watch keeps time, the better it is. Because a watch keeps time, it can be used for all sorts of things that require that the time be kept. A pocket watch can be fitted with an alarm and setting mechanism, so it can wake people up. A pocket watch can come equipped with an extra window so that it can be used to record time intervals so it can be used as a stopwatch. These functions *presuppose* that the watch keeps time. An alarm cannot wake a person up in good time for her to get ready to go to

---

[2]See for example, the essays collected under the title *Philosophy and Design* by Vermass et al. (2008).

work if it is not connected up to a some kind of clock, and time intervals cannot be recorded unless a clock of some kind is used. However, that a watch keeps time, does not presuppose that it comes equipped with an alarm and setting mechanism; pocket watches never used to have this feature (and they don't normally have it now). And keeping time does not presuppose that a watch is also a stopwatch, a watch can keep time without this feature either, and again stop watches are a relatively recent invention. These latter two functions I call *derivative* purposes because they presuppose or derive from the primary purpose of the artefact, but not vice versa. Watches and clocks are extraordinarily important and they have many more derivative purposes. Pocket watches were, and still are, decorative, especially when worn with a fob. They can become collectors' items and highly sort after. They can even be used, so it seems from the movies, as aids to hypnosis when swung beguilingly in front to the subject's eyes. But none of these uses or purposes depend on the primary purpose of telling the time, nor indeed on any derivative purpose, and for this reason I refer to them as *secondary*.

While the primary purpose of an artefact is the main and usually the exclusive focus of the designer, it is possible that she foresees that her creation has certain uses that are in no way part of or dependent on its primary purpose and it is possible that the designer intends that this be the case. Such a scenario will be unlikely. Secondary purposes are normally unforeseen by the designer, and almost always she cannot normally be held responsible for such uses.[3] However, the situation is different for the primary purpose, and possibly for derivative purposes. The primary purpose of an artefact is defined to be what it is designed to do, and this just means 'what it is intended to do'. Design is an *intentional activity*: the designer wants to achieve something, namely make an artefact that does something, so she goes about drawing up plans, making models, etc., which she believes will bring about the end she has in mind. This is what an intentional activity is: it is one in which the agent desires to perform a certain action and believes that what she is doing will realise this end, namely, in this case, to design an artefact that fulfils a given purpose.[4] The designer may or may not also intend that the derivative purposes are realised by what she designs.[5] The designer of a pocket watch may be disinterested in any subsequent use as an alarmed watch, but the designer of a pocket alarm watch who intends her work to be used for this purpose, so she must design a system that can have an alarm mechanism easily added.

The designer is therefore *responsible* for the artefact's primary purpose in the sense that her work makes it possible: the artefact enables us to do X, the designer designs the artefact in question, therefore without her work we could not do X because to do or have X we need the artefact. This is causal responsibility. What I will call *designer responsibility* is an open-ended and multi-dimensional concept

---

[3]Unless she should have foreseen them, but I set this complication aside here—see Forge (2008) for more on this.

[4]For more on intention and responsibility see Forge (2008: Chap. 5).

[5]All of Ihde's examples involve derivative purposes, see Forge (2012: 147–148) for more.

that has a number of elements and applications; and being causally responsible for an artefact is a necessary condition for designer responsibility. So for example, if the agent is commissioned to design the artefact, then she has an obligation to produce what she is paid to produce, and in this regard, she has a professional or commercial responsibility. If X has some significant social or medical purpose, say it is a vaccine made by a new technique involving novel equipment, then the designer has a duty to make sure this is a safe and reliable method. The responsibility of the designer here is more than professional or commercial responsibility, because safe and reliable vaccines have an important role to play in safeguarding people's health, and hence it is appropriate to speak of social responsibility. A designer has a legal responsibility to ensure that the artefacts she makes possible work as they should and do not malfunction and cause harm.

Our interest here is with the designer's moral responsibility: I claim that a designer has a *moral responsibility* not to deliberately design things that are *intended* to cause harm. Artefacts are supposed to help us do things that we are not able to do without them and surely we all want such things to make life better for us. If the time, effort and ingenuity of the cleverest and most creative people among us in fact makes life worst by creating harmful things, then that is wrong. This idea of designer responsibility can explored and elaborated some more—and I shall do so—but enough has been said about it for the moment to support claim that weapons design is unique and exceptional, and in fact is *contrary* to the reasonable demand that designers endeavour to work for benefit to society or at the very least do not cause harm. It follows that if someone does such a thing, designs an artefact that is harmful, then either what she has done is wrong or she must be able to give an acceptable justification of why she did so.

3.2 The Primary Purpose of Weapons Research

Weapons research, however, *aims* to produce artefacts whose primary purpose is to harm. As such, it is the only design activity whose *raison d'être* runs directly counter to the reasonable moral demands that should be placed on designers and researchers. We might then conclude immediately that weapons research is morally wrong and that no moral person should ever engage in it. But this claim might not yet be entirely convincing and we should take some more time to make sure it is been firmly established. What is clear from what was put forward above is that a designer intends that what she does achieves the primary purpose of the artefact in question. Without this assumption, design becomes a hit and miss affair, without direction or aim, and that is simply not true. So what needs to be established, first of all, is that the primary purpose of weapons design is indeed to produce the means to harm, and then it needs to be shown that producing the means to harm is indeed morally wrong. So first we need to canvass alternative descriptions or accounts of what weapons researchers aim to do.

What else could the primary purpose of weapons research be? Weapons are the means to harm, there is no doubt about that because harming is what weapons have been doing for thousands of years. But if we look at the rationale for weapons research, and everything else to do with weapons, from recruiting soldiers to use them, to spending vast sums on acquiring them, we see that this is not about plans to harm others, but about preventing others from harming us, for defence. Elsewhere, I have referred to this as the *standard rationale* for everything to do with weapons, including weapons research (for example, Forge 2018: 11–14). Here is the US Defense Advanced Projects Agency (DARPA) mission statement:

> …to maintain the technological superiority of the U.S. military and prevent technological surprise from harming our national security by sponsoring revolutionary, high-payoff research bridging the gap between fundamental discoveries and their military use. (Mission statement www.darpa.mil)

In other words, DARPA sponsors and oversees US weapons research, and 'defense' is the rationale. DARPA wants to make sure no other states can get technically superior weapons systems to those possessed by the US military and to that end it sponsors weapons research. DARPA would, one supposes, therefore endorse the proposition that the purpose of weapons research as it sees it is to *prevent harm*, by making sure the US can defend itself.

Setting these two alternative views on the primary purpose of weapons research side by side gives us

> PP1: The primary purpose of weapons research is to design new ways to harm.

> PP2: The primary purpose of weapons research is to design new ways to prevent harm.

PP1 and PP2 could be expressed in more nuanced ways, but that does not matter for the moment, as the present form captures the main idea. A decisive reason in favour on PP1 is that a weapon can harm without ever being used to prevent harm and can be used to prevent harm only because it can be used to cause harm. The three atomic bombs created by the Manhattan Project, for example, were never used to prevent harm, only cause it. Had things been different and a demonstration of the power of the bombs been used to force the Japanese government to surrender, then these weapons would, on that scenario, have been used to prevent harm. However, that outcome would have come about because the weapons could cause harm: the Japanese surrendered because of the threatened use of the bombs. But the fact that the bombs killed hundreds of thousands of Japanese civilians was not a consequence of their being able to prevent harm. So to argue in favour of PP2 over PP1 is to conflate a *derivative purpose* of a weapon with its *primary purpose*. We should substitute

> DPDef: A derivative purpose of weapons research is to defend against harm.

> DPDet: A derivative purpose of weapons research is to deter harm.

for PP2. It is possible that there are derivative purposes besides these two; if so they will not concern us here.

The fact that the prevention of harm is a derivative purpose of weapons research does not mean that DARPA intends for the weapons it makes possible to be used to harm rather than prevent harm, and it is important to acknowledge this. We have seen that someone can design an artefact because she knows that it can fulfil a derivative purpose; this may be what motivates her and what she desires. Every single weapons researcher that DARPA employs may sincerely wish that the systems they create are never used to harm. But this does not mean that they do not intend to make the means to harm. Even if a derivative purpose is the focus of concern and attention of the designer, the fact that she must make an artefact that satisfies the primary purpose cannot be escaped, avoided or somehow finessed. All the derivative purposes of an artefact presuppose that it fulfils the primary purpose, and that is true of weapons as well as every other artefact.[6]

According to common morality, if an agent foresees that she does harm even if she does not intend to do so, then she breaks a moral rule and is to be called to account and asked to justify her behaviour: intending to cause harm is not the only basis for judgements of wrongdoing. Focussing on the difference between intention and foresight, this has to do with what the agent desires. If X is the intended outcome, then this is what the agent desires, but if it is merely foreseen, then it is, or may be, a 'by-product' of the act which brings about what the agent desires and she may not have any interest therein.[7] The reason that I have argued that the primary purpose of an artefact is what the designer *intends* and does not merely foresee, even when her main interest and concern *is* with a derivative purpose, is because the primary purpose is not merely a by-product of her work, but is central and essential to it, and indeed it is *all* that the weapons researcher does. Doing weapons research is what the weapons researcher does and this involves designing the means to harm, nothing more. The weapons researcher does not, by designing a new weapons system, thereby put in place a regime of deterrence or ensure that her country is defended. Defence and deterrence will have a role to play in the justification of weapons research and will be discussed at length in the next chapters, but they do not quality as the primary purpose of weapons research.

As I have just said, deterrence and defence will figure prominently in what follows. To conclude this section, I will make some brief remarks about the former which I will amplify later on. Deterrence, as it is understood here, is a relation between two or more states, for instance, bilateral deterrence is a relationship between two states. Suppose state B believes that state A wants to engage in certain actions that it believes is against its, B's, interests, and so it puts in place measures that it believes will frustrate A's plans because now the costs to A of its actions would be greater than the expected benefit. In other words, B is trying to deter A, to

---

[6]For present purposes, of course, it is only necessary that this relationship hold for weapons.

[7]An agent is responsible for what she intends and for what she foresees will come about as a result of her action, and more besides.

stop A doing what it would otherwise do. One way for B to set up a deterrence regime, is to threaten or imply a military response to A, and to do that, B may need to acquire suitable weapons—the relationship between the United States (US) and the Union of the Soviet Socialist Republics (USSR) between 1951 and 1986 was marked by the development and acquisition of ever more terrible nuclear weapons for the ends of deterrence. Thus, weapons can play a part in deterrence, but deterrence as a relation between states depends on contingencies of the political and historical context, and as these change, the need for deterrence may fade away.

3.3   The Means Principle

There have been examples of weapons research that have *directly* harmed people. The Japanese for example conducted experiments with biological warfare agents on human subjects in China from 1938 until 1945 (Felton 2012: Chap. 1). The British used human experimental subjects to test nerve gases (Schmidt 2015: 1), and there have been other cases. But such instances are the exception, and on the whole the conduct weapons research does not in and of itself harm anyone. If we are to argue that weapons research is morally wrong, it will be necessary to establish a suitable connection between weapons research and harming. On the face of it, the connection seems obvious enough: weapons research creates new and improved means to harm and even though the weapons researcher does not (normally) actually do the harming herself, she makes it possible for others to do so. Without weapons research, no new ways of harming would be introduced into the world. What more needs to be said? I think something more does need to be said here to support the conclusion that if harming is wrong, then so is providing the means to harm.[8]

We introduced common morality with reference to one individual agent's actions and how they affected another individual, and we agreed that moral wrongdoing is in evidence when those actions are harmful. It is possible for harm to come about, not just through the actions of one person, but through the actions and activities of two or more agents, none of which is enough on its own to cause harm. In some such instances, the harm eventuating may be accidental and in others in may be planned and deliberate and here we can speak of group harming or *organised harming* which is the outcome of a group cooperating. There are organisations who use violence as a way to achieve certain ends and who contain elements or sub-groupings whose task is to commit violent acts when called on to do so. Numbered among such organisations are criminal gangs, terrorists, insurgencies and guerrilla movements, police forces and defence forces. Some of these can use force legally in the sense that any violence they commit is state-sanctioned, while

---

[8]In this section I provide an argument in favour of the means principle. I have done so elsewhere, for instance in Forge (2012, Chap. 7) and Forge (2018, Chap. 3), but the argument here is a new one. The reader who is unconvinced, or simply interested, may like to consult these references.

others use force illegally, though some may claim legitimacy.[9] For the moment I want to focus on the methods of such groups and not on their aims, though I will choose a terrorist group or insurgency for a context. Not all the members of these organisations actually commit violent acts. For example, an insurgency normally has a division of labour, with different members fulfilling different roles, some do the planning, others organise the materials and logistics, others actually commit the violent acts, and then there are those who supply or make the weapons. Skilled bomb makers have been integral parts of insurgent groups and such organisations therefore have their own weapons researchers. The following scenarios show that providing the means to harm is morally wrong.

Suppose D is a bomb designer for such a group. She designs a device for a particular target and that is all she does: the bomb-making factory makes the bomb, it is planted and set off by an active service unit, and duly kills some soldiers. There is no doubt that D, along with everyone else who had a hand in killing the soldiers, is guilty of wrongdoing. What this example shows in the first place is that moral wrongdoing is *not* restricted to one person directly harming another: if one person knowing aids and assists others to cause harm, then that person is guilty of wrongdoing. One of the ways in which one person can aid and assist another, the way that concerns us here, is if she provides the means to harm to someone who then uses it to cause harm. If we understand "provides" in this context to signify that the means, be it a bomb, gun, or some other weapon, is made available with the intention that this be used for some specific purpose, for some given target, then the conclusion that the provider does wrong is beyond dispute. This is the import of the scenario just considered: D designs a device for a particular target and the device is then constructed and used, D is guilty of wrongdoing even if she does not actually construct the device or have any further role to play after she has designed it. And she does wrong *because* she provided the means to harm by supplying directions, in the form of a design, for making a bomb. Her work is the vital first step in the process that led to the bomb going off and people being killed and injured and their property destroyed. It appears here that D is guilty of moral wrongdoing both because she is responsible for the harm caused *and* because she provided the means that were used to cause the harm. I maintain that weapons researchers are *always* morally accountable for designing the means to harm; whether they are also accountable for the actual harms caused is an open question and not something that needs to be finally decided here.

In response to this claim as it is framed in the context of the example, it could be said that what D is guilty of is the harms caused because she designed the means:

---

[9]Two comments: not all violent action by the police and the armed forces is state-sanctioned, the violence must be necessary and within certain bounds and limits. Criminal gangs, one assumes, acknowledge that what they do is illegal, while the other hand, insurgents normally believe that they trying to overthrow an illegal regime. The latter commit violence on political grounds, in contrast to the former.

the means are exactly that, just means, and the moral wrongdoing only concerns the harms caused. That she designed the means is the *ground* for the judgement of moral wrongdoing, not the content thereof. As a first reply to this objection, that designing the means is not itself morally wrong and continuing with the example, suppose the authorities uncover the plot and capture the group, prevent the bombing so no harm is actually caused, and track down D. Each member is guilty of moral wrongdoing because they belonged to a terrorist group, because the planned a bombing *and because* of the particular role they played in the group, in D's case as the bomb designer. A similar, but less convincing, response, can be made to this scenario, namely, that what is morally wrong here is belonging to the terrorist organisation and being a bomb maker is again the ground not the content of that judgement. We can reply here, as we could have done in the previous case, that wrongdoing is not 'evenly distributed' over all the members of the group, and that particular roles, like being part of active service units and being a bomb designer, are worse than others, and this suggests that being a bomb designer is wrong in itself. But there is more that we can say to put this kind of objection to rest and reaffirm our conclusion.

We now take D to be a free-lance bomb designer who sells her designs to any organisation who will pay for them. D does not know where, when or if her designs will ever actually be made or used, and she does not care. What D does is morally wrong, not because of any actual harms caused, not because she belongs to any particular group, but because the bombs could be made and could harm others, and this is morally wrong because D has provided the means to harm. If this is not enough, if it is said that it is wrong sell the means to harm, then this is just to grant the point, that what is wrong is what is sold, not that it is sold. It would be equally wrong to make bomb designs, or indeed any weapons design, freely available, for instance by posting the material online.[10] And again, what is wrong is that designing weapons provides the means for harm and hence makes harming possible in new ways. Finally taking this sequence of scenarios to the limit, if D is an eccentric inventor who spends her time designing new types of bomb but never makes these public in any way, and to avoid the risk that these designs might be stolen always destroys them, do we say that she is also guilty of moral wrongdoing? Not under the stated condition, because she does not *provide* the means to harm: she does not make a design available to anyone who could use it to harm, therefore it is not an instance of moral wrongdoing. In all the other scenarios, D makes the design *available* and that is wrong.

We can sum up all this by

MP. If it is morally wrong to harm, then it is morally wrong to provide the means to harm.

---

[10]3-d printers have made it possible to build weapons directly from digital designs by a process known as additive manufacturing. See Maddox (2013) for a discussion of the technique and what it might produce.

Elsewhere, I have referred to this idea as the Means Principle.[11] MP is a conditional and we can detach the consequent by appeal to HP, given that weapons research provides the means to harm, and assert

WRMR: Weapons research is morally wrong.

I believe we should accept MP and interpret it as applying (at least) to all episodes of weapons research, which implies that WRMR is true. There are, however, two objections to this proposal, and they both stem from the same mistake, namely a conflation of judgements of wrongdoing and about justification. The first objection, similar to one raised above, is that the conclusions drawn from the series of scenarios are only persuasive because D is portrayed in some as a member of an illegal organisation and trades on the assumption that it is wrong to belong to such an organisation. But this misses the point. The judgement that it is wrong to belong to a insurgency is based in part on the view that the aims of such groups are illegal. By definition, an insurgency aims to overthrow the government, secure rights for a minority, and so forth. These aims may or may not be just and they may or may not serve to justify the methods which the organisation uses, and, as is often the case, the received view of the matter when the conflict is over is the one held by the side that won: Fidel Castro's The Movement was a group of heroic freedom fighters whereas the Malayan National Liberation Army (which was defeated) was a terrorist organisation. Castro, and members of the IRA and the Viet Cong and other successful insurgencies, maintained that the violence they committed was justified because their cause was just. The point, however, is that violence needs to be justified because it is morally wrong: all violent acts are morally wrong regardless of who commits them, and some can be justified, others cannot. MP thus asserts that those who provide the means to harm do what is morally wrong, *regardless* of which side, group or cause the means are made available to. MP helps us decide who is accountable for violent harmful acts and hence who needs to justify what they do. Justification is the next step; it is what is needed for the *prima facie* judgement of moral wrongdoing to be withdrawn.

The second objection is that what bomb designer D does is morally wrong because she is making an offensive weapon available, and in general, weapons research that intends to produce offensive weapons is wrong because such weapons cause harm. This neglects defensive weapons, whose rationale is to prevent harm. If causing harm is wrong, then surely preventing harm is at least morally permissible? If this is right, then we will have to revise our account of the primary function of a weapon, because the claim here is that there are not only weapons that are the means to harm, but also a category of weapons that prevent harm. If this objection could be sustained, then it would not only entail a revision of the primary purpose

---

[11]See for instance, Forge (2012: 136–144). As I noted above, the argument here for MP is a new one. Also, the statement of the principle in Forge (2012) includes the qualification "without justification". Seumas Miller, in his book on dual use, introduces a similar principle, which I discuss in the appendix on dual use.

of weapons, but also of MP. MP would only apply to offensive weapons. We will see that there is in fact no separate category of weapons that can only be used to defend against harm and hence that the objection cannot be sustained. It will be necessary to look into the issues in some depth and this will be done in the next chapter. Defence, as we shall see, is the main, indeed only, justification for violence and harming: defending oneself and others from harm is the only justification for harming. If there were weapons that *only* prevented harm, then weapons researchers would not themselves need to justify what they do—given that their work was intended to produce defensive weapons—as I believe they must, based upon the argument of this section which supports MP. They would not need to do so because they would have an excuse: we are not providing the means to harm, but the means to prevent harm, therefore there has been no wrongdoing and we do not to account for our actions.

3.4 Conclusion

In this chapter I claim to have provided sufficient reasons for us to accept WRMR, that weapons research is morally wrong. I have argued that weapons research provides the means to harm, when it provides designs for weapons. Designs are plans, instructions, know-how, etc. which those with the requisite skills and resources are able to produce the artefact in question, the artefact that has been designed. Artefacts can be used for a variety of purposes, but the focus of attention of the designer is what I have called the primary purpose. It follows that designers are 'committed to' and hence responsible for the artefact being used for its primary purpose. This is what I called designer responsibility. I argued that the primary purpose of weapons is to harm, to kill and to destroy. To conclude that weapons research is morally wrong it was necessary to establish MP, which was done in the last section of this chapter. This concludes the first step in the case against weapons research.

Appendix: Dual Use and Weapons Research

Some ten years ago I wrote a short note attempting to clarify the notion of dual use, a topic that had at the time been attracting interest and which continues to do so. I distinguished three different categories of dual use items: research, technologies and artefacts. These are clearly different sorts of things. Research is an activity, while technology, as we have seen, is a form of knowledge, knowledge of the techniques for the production of artefacts, whereas artefacts are objects. Certain instances of these kinds are classed as dual use because they can have both a 'good' and 'bad' use. Ammonium nitrate, for example, is normally used as a fertilizer, but it can also be used to make so-called fertilizer bombs.[12] The dual use categories are related: research aims to give us technology, which in turn produces artefacts. It is the latter that normally have the immediate impact on the individual, be this good or bad. My suggested definition of a dual use item was as follows:

> An item (knowledge, technology, artefact) is dual use if there is a (sufficiently high) risk that it can be used to design or produce a weapon, or if there is a (sufficiently great) threat that it can be used as an improvised weapon, where in neither case is weapons development the intended or primary purpose. (Forge 2010: 117)

Threats differ from risks here in that they signify the intention to do something harmful, rather than the possibility that something harmful might happen. I have coupled risk with the design of (new) weapons but threats with improvised weapons. States threaten one another, as Iran and Syria are doing at present, though the threat here is usually conditional. However, in regard to dual use, I think we are more concerned at the moment with sub-state actors, such as terrorist organisations, using or adapting existing items for improvised weapons, as has been done with ammonium nitrate, than with states arming themselves with improvised weapons, as states do not usually need to improvise weaponry.

So much by way of background. There have been a number of issues about dual use that have been discussed in the literature, but what is of interest here is whether undertaking dual use research is itself morally wrong, or whether it is morally permissible—taking advantage of dual use research and developing weapons is morally wrong. In the second place, and this is not something discussed elsewhere, are we able to express the notion of dual use with reference to our taxonomy of the purposes of artefacts and is it helpful to do so? There are examples of dual use research in a number of scientific fields, but I will choose one from nuclear physics which took place during the Manhattan Project, namely research which enabled the technologies which produced the fissile material for the bombs. This research also laid the foundation for civilian reactor technology: in fact, Enrico Fermi built the first nuclear reactor to verify that a chain reaction was possible in 1942, and nuclear reactors were built in 1944 to produce plutonium (Forge 2012: 85). The difference between what goes on when an atomic bomb explodes and when a nuclear power reactor operates is that in the former case the chain reaction is uncontrolled and energy is released until the assembly blows apart, while in the latter the energy release is gradual and controlled, heating up water to drive turbines for example. Naturally-occurring uranium will not suffice for either purpose and must be enriched. The science and technology of enrichment, and under certain circumstances the artefacts produced, are dual use. It is necessary to develop the example a little more to see why this is.

Recall that naturally occurring uranium comes in two isotopes: chemically identical species of atom that have different physical properties owing to the

---

[12]Ammonium nitrate is strictly speaking a substance not an artefact because it does occur naturally: I am assuming here that it has been manufactured and is in that sense an artefact.

presence of differing numbers of neutrons. In the case of uranium, the lesser abundant U-235 is much more fissile, much more apt to fission and release energy, than the more abundant U-238. A sample of uranium metal is said to be enriched if the amount of U-235 is greater than 0.7%, which is the proportion found in nature. The greater the enrichment, the greater the neutron flux, the greater the flow of neutrons in the metal, and the greater the value of the neutron multiplication factor k (see Chap. 2). It is possible in theory to make a nuclear weapon with uranium enriched to a little less than 20% but the accepted practical limit is set at 20%, what is referred to as highly enriched uranium or HEU, when 30 kg of the metal would be needed (Glaser 2017: 8). By contrast, only 2.3 kg of 95% enriched uranium is needed for the same purpose—uranium enriched to 90% or greater is known as military or weapons grade. Civilian power reactors, on the other hand, only require enrichment to about 4%. A civilian power reactor fuel rod is therefore not itself a dual use item because it cannot be used to make a bomb, unless subject to further enrichment. Weapons grade uranium is not suitable for power reactors because the extremely high neutron flux makes bomb making easy but cooling very difficult, so it is not itself a dual use item either. However, there is another sort of reactor, the research reactor, which does require HEU, and hence a sample of HEU is dual a use artefact: it can be used for research or bombs.[13]

Suppose a 25 kg sample N of uranium enriched to a degree greater than 20%, say 26%, is required and the necessary means to produce N set up and calibrated, and N is duly obtained.[14] N is a dual use artefact because it can be used to fashion fuel rods for a research reactor, or, with some ingenuity, to make an atomic bomb. Let us ask: does N have a primary purpose and if so, what is it? The primary purpose cannot be as material for a research reactor or as material for a bomb: if N is fashioned into fuel rods and functions as such, this is not because it is material for a bomb, and if it is made into a bomb, this is not because it is material for fuel rods for a research reactor. The relation is not the same as it is between weapons as the means to harm and weapons as the means for deterrence. There is an alternative possibility. N will have a certain neutron flux, characteristic of 26% enrichment, so consider this to be the primary purpose of the artefact, to have this property. In which case both being material for fuel rods and for bombs stand as derivative purposes. Clearly, having a neutron flux of this value does not presuppose that it can be used for either fuel or for a bomb, but having a neutron flux of this (order of) magnitude is necessary for either function. Now we can ask if this is a useful way to talk about dual use, in terms of primary and derivative purposes.

I believe there are some advantages in so doing. In the first place, it provides an explication of "dual use": thus an instance of dual use comprises two functional

---

[13]See National Academies of Sciences, Engineering and Medicine (2016) Chap. 3 for civilian research uses of HEU. As the title of this publication, *Reducing the Use of Highly Enriched Uranium in Civilian Research Reactors*, implies, there are moves to try to reduce or eliminate HEU in research reactors.

[14]This could, for instance, be achieved by using calutrons, mass spectrometers that separate substances by taking advantage of their different masses.

elements that derive from the same primary element, which is to say that the each 'dual use' is a derivative purpose. Research can be said to be dual use as a consequence of the primary purpose of its outcome, which in turn enables the derivative, dual use, purposes. In this way we can offer an explanation of why dual use is possible. In our example, and in accordance with the definition given above, the use of N as weapons material is the bad use and as fuel rods for a research reactor the good one. We have seen that the designer is committed to the primary purpose of the artefact that she enables in the sense that this is what the artefact does, but that her motivation can come from the desire to realise a derivative purpose—for example nuclear research for reactor fuel. Dual use is problematic because research that leads to a good use can also lead to a bad use, and the assumption is that the research is motivated by the good use. This is therefore compatible with our views about the motives and intentions of the designer.

Turning to the issue of the morality of dual use research, of research motivated by the good use, I will begin by looking at Seumas Miller's recent contribution (Miller 2018). Miller also sees the problem with dual use research in terms of weapons development, and gives a number of examples, including one from the nuclear industry. The weapons he has in mind are all weapons of mass destruction, which raises the stakes beyond the possibility of IEDs made with ammonium nitrate. Miller formulates a principle which forbids this kind of weapons research: he calls this the No Means to Harm (NMH). He says "Roughly speaking, this is the principle not to provide malevolent persons with the means to harm; a principle which is ultimately derived from the more basic principle: Do no harm" (Miller 2018: 13). NMH is in the 'spirit' of WRMR, though I would have liked to see his derivation from the harm principle, which we saw is not particularly easy or straight-forward. Like many others, Miller sees dual use research as giving rise to dilemmas, as follows:

> Option 1-Scientists morally ought to conduct research into nuclear fission since it enables the provision of a much needed source of power for civilian purposes.
>
> Option 2-Scientists morally ought not to conduct research into nuclear fission since it led to the creation of the atomic bomb and, ultimately, nuclear weapons capable of destroying humanity (Miller 2018: 12).

Option 2 could be better expressed, though one gets the point. But what is the basis of the obligation to work on the beneficial outcome, is there a principle which obliges scientists to prevent harm or convey benefit?

If there were no such principle, nor any other reason to work on the beneficial aspect of the project, then there would be no dual use moral dilemma: the research would be proscribed on the basis of NMH and there would be no contending principle that promotes and justifies the work. This is precisely the implication of my account of the matter: I see no dilemmas here at all. This, of course, is a consequence of the moral framework adopted in this book which does not prescribe any action that is supposed to bring about good outcomes. I note that Miller does not put forward any such principle either. It is possible to adopt a different moral framework which does oblige people to bring about good outcomes as well as avoid bad ones, such as CP, and dual use dilemmas will only arise in such a context. Dual use research is not weapons research so it is not a main focus of interest here. But the position put forward in Chap. 3 implies that there is a sound moral reason not to conduct dual use research, because it risks weapons development, and since there is no other reason to undertake such work, no moral rule, the present account implies that dual use research is morally wrong.

Appendix: Lethal Autonomous Weapons and Designer Responsibility

There are, as far as we know, no lethal autonomous weapons (LAWS) in existence at present. One can speculate about just what kind of things LAWS will be. I will not go down this path, but just think of LAWS on analogy with drones, drones that are capable of choosing and carrying out missions autonomously, without a pilot.[15] Such systems will have to have learning capabilities and flexible programming. This category of weapon is of current interest, and there have been appeals to ban the research and development of LAWS because of concerns specific to this par-ticular kind of weapon. As far as we are concerned, WRMR applies to LAWs because they are weapons and our principle applies to all weapons research, so no one should design such weapons, and we can agree with those who call for bans on any research into LAWS. In this regard there is nothing special about LAWS. It is nevertheless worth discussing LAWS here because of one of the reasons why bans have been called for is the possibility that when LAWS are used no one would be responsible for what they do, especially if they make mistakes and kill the wrong targets—there may, it is alleged, be *responsibility gaps*. If mistakes mean that the weapons has malfunctioned, then some like Robert Sparrow think that the designer is an "obvious and important possible endpoint for the allocation of responsibility" (Sparrow 2007: 178). I am inclined to agree with him, but there is a little more we can add to the discussion on the basis of our account of designer responsibility.

Much of the discussion around this issue, and about LAWS more generally, has focussed on autonomy and the senses and ways in which LAWS could be auton-omous. Gubrud has suggested that landmines represent a good starting point for talking about LAWS because they are autonomous in the sense of not being

---

[15]At the time of writing, a group of government experts is meeting to discuss LAWS, Reaching Critical Will (2018). One of the most important tasks, before any attempts can be made to limit their development, is to characterise the systems, to say just what LAWS are. Without making an attempt to summarise the contributions by both governments—included here are the US, Russia, China, the UK and France—and non-government organisations, a recurrent theme is that 'genuine' LAWS would be able not only to engage targets autonomously be also select them autonomously.

triggered by their operator, and there have been other useful contributions.[16] For the present purpose it is not necessary to canvas these alternatives. It is sufficient to assume that a weapon is autonomous if it *selects* its targets as well as launching whatever ordnance it is equipped with. In other words, something is a LAW if it decides who to kill. So for instance, it could be something that functions in the same way as the current generation of drones, in every way except that there is no pilot, no sensor analyst and no intelligence officer (see Gusterson 2016: 33). If a drone killed the wrong people, then it might be that the pilot, the intelligence officer, or the sensor operator would be called to account, but these persons are 'out of the loop' when it comes to LAWS.

The following statement by Andreas Matthias, who raised the problem of responsibility gaps and has been widely quoted, is a summary of a general problem with advanced AI systems, systems that have the kind of learning capacity and flexible programming that we associate with LAWS:

> Traditionally, the operator/designer of a machine is held (morally and legally) responsible for its operation. Autonomous learning machines ... create a situation where the designer is in principle not capable of predicting the future machine behaviour and thus cannot be held morally responsible for it. The society must decide between not using this kind of machine any more ... or facing a responsibility gap which cannot be bridged by traditional concepts of responsibility ascription (Matthias 2004: 175).

I left as an open question the responsibility of the designer for the harms caused when her weapon is actually used. Thus in the case of the bomb designer D, I said that if she designed a bomb for a specific target which was then used for that end, then she is responsible for the harm caused. Similarly, if she knew that a bombing campaign was planned and designed bombs for the group, but did not know anything about the specific targets, then we should also hold her responsible for the harms. It would seem however as the harming events become more remote from the time and place where the bomb design takes place, it becomes increasingly difficult to hold D responsible for the harms caused as well as for designing bombs. This accords with the received view of responsibility.

In general, if agent P does not know about the outcomes or consequences of her action when she performs the action, then it may appear unfair if she is held accountable for them. So it seems we should accept: P did not know that by doing X, Y would come about, therefore P is not morally responsible for Y and so cannot be held to account if Y is harmful—this (Aristotelian condition) is implicit in the passage from Matthias. I argued in my book *The Responsible Scientist* that ignorance does not always excuse and that we need a 'wide view' of responsibility in which an agent who is in a 'position to know' about the consequences of her action, and who should have known what these are, can be held to account. If P is a

---

[16]Gubrud refers specifically to the Ottawa convention banning antipersonnel landmines, which defines them as "'Anti-Personnel mine' means a mine designed to be exploded by presence, proximity or contact by persons and that will injure, incapacitate or kill one or more persons" (Gubrud 2018: 1).

designer and what she does is design an artefact, which is then used as it is supposed to be used, then evidently she is in a position to know this. That is to say, she is in a position to know that what she designs will work or function in the way she intended it to be used, given that she designed it properly, even if she was not in a position to know all its particular uses and applications, stretching into the future. For virtually all artefacts, such issues about responsibility and accountability will not be either interesting or important. Matters are different when it comes to weapons research.

Expressing the issue in the light of this account of responsibility, if the designer of a LAW has no reason to suppose that her weapon will 'malfunction' by selecting and killing the wrong target, can she be held responsible? To answer this question, we need to ask first who is responsible when the LAW selects and kills the 'right' target, those who it has been flexibly programmed to seek out and destroy. And here the issues are the same as they are when weapons are used that require human decision-making to select targets, including the soldiers, those in the chain of command, and those who provide support and logistics for the mission in question. The weapons designer is also responsible if she was in a position to know about the mission. In regard to LAWS, we can speculate about how this might be the case: suppose the mission is to target insurgents in the tribal homelands of Pakistan, where the majority of drones are now employed. The designer of the LAWS needs to be aware of this because it sets some of the flexible programming parameters— LAWS need to know where they are. In which case the designer is responsible for the harms her creations cause, *regardless* of who is killed, whether or not the right targets have been destroyed, because in the first instance it is always wrong to harm. The next question is whether or not the harming is *justified*, and the assumption is that if innocents are killed—the 'wrong targets'—then it is not but if insurgents are killed, then it is. There is room for dispute and debate, but this is not the issue here.

It is the situation that is the converse of that just considered that is the interesting one: suppose the designer is not in a position to know about a given mission, and so would not according our account of the matter be held responsible for the harm caused in the normal course of events, but now the wrong targets are selected and engaged and innocents are killed. Is the designer responsible? There are two approaches possible here, one of which appears to give rise to responsibility gaps and one that does not. The first sees the LAW as analogous to a soldier who kills an innocent person. If his commander had no reason to think he was not properly trained, subject to stress, or that there was anything else to impair his ability to carry out his mission, then the responsibility for killing the wrong person rests squarely with the soldier, he is accountable. If the soldier is replaced by a LAW in a comparable situation, then it looks as if no one, or nothing, can be held to account, and there is a 'responsibility gap'. The second approach sees the LAW on analogy with the weapon the soldier uses. Suppose this is some kind of missile with a guidance system and it goes astray and hits the wrong target. If the system is faulty and the problem is not the soldier's aim, then the question of the designer's responsibility is relevant. And the same is true for a LAW. If these engage and kill the wrong targets, then they are faulty, they do not work as they should and the designer is accountable. There is no responsibility gap. I opt for this second option and hence I endorse Sparrow's view of the matter and I think Matthias is wrong in his assessment of responsibility for autonomous learning machines when these malfunction, at least when it comes to LAWS.

References

Felton, M. 2012. *The Devil's Doctors*. Barnsley, Yorks: Pen and Sword Books.
Forge, J. 2008. *The Responsible Scientist*. Pittsburgh: Pittsburgh University Press.
Forge, J. 2010. A Note on the Definition of Dual-Use. *Science and Engineering Ethics* 16: 111–118.
Forge, J. 2012. *Designed to Kill: The Case Against Weapons Research*. Dordrecht: Springer.
Forge, J. 2018. *The Morality of Weapons Design and Development*. Hershey, Penn.: IGI.
Glaser, A. 2017. *Highly Enriched Uranium, Research Reactors and the Risk of Nuclear Proliferation*. Washington: American Physical Society.
Gubrud, M. 2018. *The Ottawa Definition of Landmines as a Start to Defining LAWS*. https://autonomousweapons.org/the-ottawa-definition-of-landmines-as-a-start-to-defining-laws/.
Gusterson, H. 2016. *Drone: Remote Control Warfare*. Cambridge, Mass.: MIT Press.
Maddox, J. 2013. Additive Manufacturing and its Implications for Military Ethics. *Journal of Military Ethics* 12 (3): 225–234.
Matthias, A. 2004. The Responsibility Gap: Ascribing Responsibility for the Actions of Learning Automata. *Ethics and Information Technology* 6: 174–183.
Miller, S. 2018. *Dual Use Science and Technology and Weapons of Mass Destruction*. Dordrecht: Springer.
National Academies of Sciences, Engineering and Medicine. 2016. *Reducing the Use of Highly Enriched Uranium in Civilian Research Reactors*. Washington: The National Academies Press.
Reaching Critical Will. 2018. *Documents from the 2018 CCW Group of Government Experts on Lethal Autonomous Weapons Systems*. http://www.reachingcriticalwill.org/disarmament-fora/ccw/2018/laws.
Schmit, U. 2015. *Secret Science*. Oxford: Oxford University Press.
Sparrow, R. 2007. Killer Robots. *Journal of Applied Philosophy* 24: 66–77.
Vermass, P., et al. 2008. *Philosophy and Design*. Dordrecht: Springer.

Chapter 4
Defence and Deterrence

4.1 Introductory Remarks

The primary purpose of weapons is to harm: this is what weapons do, this is what they are designed to do, and the more effectively and efficiently they harm, the better they are as weapons. Weapons are exceptional in this regard, for no other artefacts are intentionally produced to do something that all of us agree is bad.[1] If this is so, then there must be compelling reasons why weapons are made, why people design them and manufacture them. If weapons harm us, why have them? And there is only one plausible answer: we must have weapons to *prevent* harm. Harms can be prevented by resistance, by defence, and by making the cost of harming prohibitive, by deterrence. Weapons have a vital role to play in both of these preventative measures, for if weapons are the most effective ways to cause harm, then are they not the best means to prevent harm? This is not controversial and it is not surprising, because such claims are exactly what we might expect, and are widely, if not universally, accepted. States, for instance, refer to their armed forces as the "defence force", governments justify spending vast sums on weapons by invoking defence, deterrence, security, and so on, and hardly anyone objects. I refer to all such views, beliefs, justifications, etc., as the 'standard rationale', for weapons research and for everything else to do with the acquisition of weapons. This is not to say that some states have not developed their military forces with a view to waging aggressive wars, but for a hundred years or so, very few if any have admitted to this openly and in advance.[2]

---

[1]Harming can of course be justified, depending on the circumstances or context, in which case, although all harm can be regarded as bad, not all harming is wrong. But weapons do what they do regardless of the particular features of the context into which they are introduced, given that what they do is harm. This warrants the unqualified judgement, that all of us can agree with, that intentionally producing something whose essential nature is to harm is a bad thing to do.

[2]As I have noted before, Hitler's armed forces were referred to collectively as the Wehrmacht, which means defence force.

J. Forge, *The Morality of Weapons Research*, SpringerBriefs in Ethics,
https://doi.org/10.1007/978-3-030-16860-5_4

In the last chapter I argued that the primary purpose of weapons is to harm, weapons are the *means* to harm, and hence that weapons research is morally wrong. This leaves open the possibility that there is weapons research that aims to create defensive weapons, and so defend against, and hence prevent, harm. This would seem to be permissible: MP would not apply to this kind of weapons research, and there would be no basis on which to judge that defensive weapons research is morally wrong. But if it can be established that defence a derivative purpose, as I claimed is case for deterrence, then this *presupposes* that weapons research that aims to produce the means to defend must, in order to do so, produce the means to harm. Defence as a derivative function of weapons presupposes its primary purpose, that it is a means to harm. If follows that there is one, not two, kinds of weapons research and hence that there is no separate category of defensive weapons research. Assuming that this can be done, and that is the task of the first part of this chapter, the focus of our attention will turn to defence and deterrence as *justifications* of weapons research: weapons research is not permissible without adequate justification.

I have said that the prevention of harm is the standard rationale for everything to do with the acquisition of weapons, including weapons research, and so it might now be asked why that issue was not addressed directly: why worry at all about primary purposes, etc.? One response has to do with the issue raised at the end of the last chapter, with the suggestion that some weapons are (exclusively) the means to prevent harm. If that were true, then for such 'inherently defensive' weapons research, no justification would be necessary. For this reason it is important to be clear on the primary purpose of weapons. It is also necessary to be clear on the relationship between the various purposes of weapons, and to keep clearly in mind that deterrence and (as we will see) defence are derivative purposes. Most importantly, granted that it has been established that weapons research is morally wrong, then moral persons are obliged to justify their participation in the activity. And they must do so with reference to the harms prevented by their work, in terms of deterrence and defence. That demand only comes into clear focus once the relationship between the various purposes of weapons has been sorted out and when weapons research is seen to be morally wrong and hence in need of justification. In this chapter I will show that there are no weapons that cannot aid aggressive war, wars that aim to take by force assets that rightfully belong to others or to infringe the liberties of others, or to harm them in other ways.

Wars and other forms of organised violence are normally conducted by states, and wars of aggression are normally wars of conquest in which one state seeks to take land and resources from another. Just War Theory (JWT) asserts that self-defence and other-defence are the only just causes for war, which is to say that armed resistance against aggression is the only just cause for war—to say that a war is just means simply that it is justified and hence morally permissible. Bad things happen in wars, as we all know, and if doing those bad things is not to commit wrongdoing, the war itself must be just.[3] In Chap. 5 I discuss JWT in more detail and mention the other conditions required for a war to be just. The aim of this chapter is to say what defence is, and see whether there are weapons that can only be used for defence, something that needs to be done before we are finally convinced that justification for weapons research is needed, and in order to understand what defence as a justification amounts to. But defence seems to be quite a straightforward idea. Referring back to HP, defence can be described as the protection of interests against others 'setting them back', and a defensive weapon is one that can only be used for this end. And in regard to deterrence, we have seen that deterrence is a state of affairs in which one party has determined that its costs, the total set backs of its interests, would be more that it would gain by compromising the interests of another party, and so desists from aggression. I discuss deterrence again in the final section of this chapter. Throughout the chapter it is necessary to give examples of types of weapons, engagements, etc., mostly from the two world wars, and for deterrence from the Cold War.

4.2 What is Defence?

In this context—when we are talking about weapons, wars and so on—it is customary to talk about *assets* rather that interests, so let us say that defence is the protection of assets. A wide range of things can be viewed as assets, from those possessed by individuals to those that are held in common, such as parks, infrastructure, and land. I will include as individual assets all those things that people have an interest in, such as their lives and well-being (though referring to one's life as an asset is a little strange). Some assets are concrete while others are not; institutions, culture, even ways of life can count as assets. Indeed, it seems that anything that has value or is valued qualifies as an asset. To take or destroy certain kinds of assets presupposes that others are also destroyed or otherwise compromised. In order to destroy a people's way of life, for example, it is usually necessary to occupy their country, and conversely, to protect a way of life from that kind of aggression is to resist such invasion.[4] Here it will be enough to focus on the defence of physical assets, of things that are located in some place that can be defended. Some of those things are military assets, the very things that are the means for defence, which can in turn be the object of defensive measures.

Defence can be active or passive. Passive defence involves putting assets somewhere that is either hard to find or hard to break into. Fortification is a method

---

[3]I will state here that I have no commitment to (any version of) JWT, and I don't think it has much to do with real wars. However it is useful way to raise issues about weapons research. For example, I will argue that if weapons research is conducted to prosecute a war, it renders that war unjust. A surprising conclusion.

[4]It was in the colonial period, from the 16th to the 19th centuries, that ways of life and cultures were routinely destroyed, by the Spanish and British in particular.

of passive defence, and fortified places have been in existence for some ten millennia—there is evidence that the city of Jericho had a wall round it in the eighth millennia BCE (Kenyon 1960, *passim*). There are lots of ways to build fortifications, but walls and moats are the classic forms. A fortified town can be attacked, or laid siege to, in three ways: by camping outside the walls and starving out the population or by assault on the walls, or both. The former is a kind of passive offensive, the aim being to deny the population the supplies they need to survive and so force them to surrender. The attackers need to be sufficiently provisioned for this purpose, so the contest becomes one of attrition. If the attackers have the means, then they may seek to reduce the fortifications by bombardment or mining, techniques that came into their own with the advent of gunpowder, or even climb over or under them. The earliest evidence of artillery are missiles that were found near the Sicilian city of Motya dating from the very beginning of the fourth millennium BCE which are believed to have been used to shoot defenders on the city walls. The catapult artillery described in Chap. 2 was first developed as siege artillery. Building walls and other fortifications does not, however, count as weapons research.

Active defence engages the attackers forces by fighting back, and so protecting the asset. It was also mentioned in Chap. 2 that the catapult was adapted for defence, by Archimedes in Syracuse among others. Early in the gunpowder age, when cannon had become highly effective against the traditional high curtain wall of fortified towns, provisions for active defensive measures, in the form of artillery towers, were incorporated into new sorts of fortifications, collectively known as the trace italienne. The coming of 'military architecture' represents a blurring of the distinction between active and passive defence: when towers, embrasures, roundels, etc., were specifically made to improve the effectiveness of cannon, then this counts as weapons research if its aim is to design a weapons' platform. There may be cases for which it is difficult to decide whether the measures are active or passive, or mixed, and where it is hard to determine whether there is weapons research going on, but that does not matter here because we do not need to make any sharp distinction between kinds of defence. In the nineteenth century, powerful accurate long-range artillery firing high explosive shells meant that towns could no longer be defended by passive measures, and so active defence became the only option. Towns and cities are now completely vulnerable to attack by ballistic missiles, for which there is no defence at all.

4.3 What is, or Could Be, a Defensive Weapon?

In the jargon of the military, a particular task is called a 'mission', and is characterised by its objective. A defensive mission is therefore a task whose objective is the protection of an asset. We could start our inquiry by focussing on this simple kind of military endeavour, and ask what sorts of weapons should be used. To answer that question, one would need to know what the asset to be protected is, where it is, how important it is, and what sort of threat is being, or could be, mounted against it. There are all sorts of possibilities here, perhaps corresponding to all sorts of defensive weapons, so perhaps this is not after all a good way to begin. A defensive mission may be local and circumscribed, but at the other extreme, there are defensive wars and these may be worldwide and long-lasting. Defensive wars may (well) require more resources than defensive missions, and so if attention is directed to the former rather than the latter and the assumption made that anything that aids a defensive war is a defensive weapon, it seems likely that a much wider range of weapons will fall into the defensive category. Perhaps what it takes to fight defensive wars will be pretty much the same as what it takes to wage aggressive wars, in which case there will be *no* special category of weapons that are properly speaking defensive weapons.[5] And even if there were such a category, if it turns out that defensive missions are also necessary for offensive wars, then the weapons that are optimal for that purpose would not be *inherently* defensive.

At this point we should pause and remind ourselves of exactly what is at issue. We are looking for weapons whose *only* role is to prevent harm by protecting assets: if there are such weapons then their function is fixed in advance and independent of the context in which they are used. If the only thing weapon w can do is protect an asset, then, by fiat, it cannot attack an asset. Let us call w a *purely defensive* weapon. The suggestion is that weapons research directed to designing purely defensive weapons is morally permissible because such weapons only prevent harm or only cause harm in order to prevent harm, and we began this section by setting out some ideas about how we might find some weapons that fit into this category. For the remainder of this section I will give reasons why weapons like w are hard, if not impossible, to find. But even if they were plentiful, we should still reject the suggestion that weapons research directed to designing weapons of this kind is morally permissible because, as we shall see, offensive wars require defensive missions. I will argue for that proposition in the next section.

Turning to some specific examples, the two world wars of the last century were wars of aggression and conquest by Germany and its allies, while Britain, France and Russia/The Soviet Union were the victims, with France and the Soviet Union being two of the countries Germany invaded. It can therefore be said that the wars were defensive from the point of view of Britain, France and the Soviet Union. The weapons that were used primarily in the First World War, the ones that caused the overwhelming majority of casualties, were machine guns and heavy artillery. The defence held sway in that war, with stalemate on the Western Front for four long years. Since heavy artillery and machine guns were the weapons that maintained that state of affairs, then are they not therefore defensive weapons? But *both* sides had heavy artillery and machines guns, the aggressors *and* the resistors of aggression—not only did they have the same kinds of weapons, the artillery was all

---

[5]The more perceptive writers on war and weapons agree. Martin van Creveld, for instance, writes "As the Arab [the aggressors] use of antiaircraft missiles during the 1973 war against Israel has demonstrated once again, the distinction between 'offensive' and 'defensive' weapons is largely spurious" (van Creveld 1991: 177).

based on the same series of innovations from 1870 to 1895 and the machine guns were all variants on the Maxim gun.

Artillery was also used by Germany in its invasion of Belgium at the beginning of the First World War, in particular in the Battle of Liege where huge German guns reduced the city's large fort complex in less that ten days, and so artillery is not a purely defensive weapons (and neither are machine guns). These weapons *became* defensive when combined with a new static defensive arrangement, namely defence in depth comprising networks of trenches and barbed wire, manned by lots of infantry. Walls and forts could not resist heavy artillery, and had not been able to do so since about 1870, but entrenched soldiers could, though at great cost. Since artillery and machine guns became emblematic defensive weapons when combined with this (passive and static) defensive structure, perhaps the status of a weapon as defensive depends on what else is available, what other military technologies are around, and how these are integrated? I believe that this is the correct account, that a weapon is used defensively or has a primarily defensive role, in virtue of what else is available, both in the form of other weapons, systems of static and passive defence and more generally on military strategy and doctrine, and even geography. I will give some more examples as evidence in favour of this claim. If it is the correct view, then there will no purely defensive weapons: being defensive is *always* contextual, *always* a matter of time and place.

After the First World War, a consuming topic for (the small cadre of) military analysts was what kind of offensive weapons, tactics, etc., would be able to break the defensive stalemate of trench warfare, were there to be another European war. An (obvious) answer was given by the English strategist Basil Liddell Hart, shortly after the end of the war: *mobile* forces are what is required.[6] When confronted with a heavily defended position comprising systems of trenches designed to repel a frontal assault, the sensible thing to do would be go round it and attack from the flanks or the rear. The means to do that were not available in 1914. Soldiers arrived at the front on trains and had to walk everywhere, with horses used to tow guns, supplies, the wounded and everything that soldiers could not carry. Flanking moves were therefore always too slow and more trenches could be dug to block the way. But soldiers in trucks—motorised infantry—artillery pulled by trucks spearheaded by tanks represent mobile forces, which could out-manoeuvre static positions, and made even more effective if in concert with airplanes.

Mobility was Liddell Hart's answer to the problem, and his ideas appeared to be confirmed some twenty years later when Germany overran France's static Maginot Line, which had been built after the First World War to frustrate any future German invasion, in a matter of weeks with mobile forces. The Germans pursued the same *blitzkrieg* strategy in 1941, with the invasion of the Soviet Union. If mobility is the key to attack and invasion, then mobile forces like tanks, motorised infantry, field artillery, airplanes, etc., could be classified as offensive weapons. If there were purely offensive weapons, weapons that could only be used for aggression, then this

---

[6]For a discussion of Liddell Hart's ideas, and for more examples, see Forge (2012: 161–166).

would seen to cast doubt on the claim made above about weapons having a defensive role being dependent on what else is available. Why is defence relative in this way and offence not? The same is, however, true of offensive weapons and we already have one example: German heavy artillery were offensive weapons in the Battle for Liege, in part because the German forces were not confronted with a massive defensive setup comprises trenches and reinforced bunkers—they were in 1940, by the Maginot Line but they knew how to bypass it.

The importance of mobile forces was recognised not only in Germany, but also in the Soviet Union. That country had some innovative military thinkers, such as Mikhail Tuchachevsky who helped to formulate the "Deep Battle" doctrine, but many were killed by Stalin in the great army purges of 1937–1939, and that nearly caused the country to be defeated by Germany in 1941 (Lee 2016: 419–424). However, they had made and designed lots of very good tanks in the 1930s as a response to ideas about the importance of mobile forces. The key to all the fighting on the vast Eastern Front between 1941 and 1945 was mobility. The German army reached the gates of Moscow in December 1941 primarily because of its mobile panzer (tank) divisions.[7] That was the high watermark of the German advance in the Soviet Union, but geographical factors—the size of the country, the lack of road and rail infrastructure, and above all the weather—prevented further gains, as did the remarkable Soviet resistance. The Soviets were able to handle the conditions of the country better, and their tank armies gained the upper hand in December 1942, after Stalingrad and finally in August 1943 after the battles around Kursk. One could examine these two massive engagements in detail and describe the weapons, defensive arrangements and tactics, but what would become apparent is that the protagonists had the same kinds of weapons and used them both for attacking and defending, depending on who had the initiative, what the ground was like, what the weather was like, and so on. There were no weapons reserved only for defence or only for attack.

4.4 The Levels of Strategy

I said that even if there were purely defensive weapons, this would still not be enough to show that weapons research intended to design such weapons is morally permissible. The reader may think that is why I have not made a big effort to find

---

[7]As an example, consider the history of one of Germany's elite World War Two mobile formations, the 7th Panzer (tank) division. This division fought from 1939 to 1945, throughout the whole war with periods for rest and refit. It was involved in the invasions of Poland, France and the Soviet Union, and until the tide of the war changed after the Battle of Stalingrad in 1942, it was part of a force that was offensively 'postured'. After Stalingrad, Germany was on the defensive in the East and fought for three years to delay the Soviet invasion of Germany. The same formation, the 7th Panzer Division, was thus an instrument for both offence and defence. See Stolfi (2014) for more.

more examples of purely defensive weapons: I suggested a hypothesis that would explain why there are no such weapons and I gave some examples to support it, but I did not consider any modern up-to-date systems—what, for instance, of anti-ballistic missile (ABM) systems, are they not purely defensive? And in any case, if it does not really matter in the end if there are any purely defensive weapons, why discuss them at all? These are all reasonable questions. To begin with the last one, to see why it is not morally permissible even to work on purely defensive weapons, we need to know something about defence and about weapons that can be used in defensive missions, and those that seem to be purely defensive. Moreover, given the standard rationale for all aspects of weapons acquisition, it is a good idea to be able to show convincingly that defence is not some kind of 'independent variable', which prevents harm regardless of any other considerations. As to ABM systems, and ballistic missiles, I will have something to say about these weapons in the appendix to this chapter.

Defence as the prevention of harm, or of causing harm only to prevent harm, is assumed to be a good thing, because if harming is bad, then preventing harm must be good, or at least permissible, as I have said. And as we saw in Chap. 1, not all harming is morally wrong: to set back an interest that is not rightfully held is not to do wrong. If I aid a runaway slave, I set back the interest of the slave owner but I do not do the wrong thing. Continuing with this example and complicating it a bit, suppose I become aware that a group of abolitionists is intending to intercept and probably injure slave catchers in pursuit of a runaway slave. I am in a position to prevent the abolitionists intercepting and harming the slave catchers, and I do so. It is doubtful, to say the least, that my action was morally permissible. The abolitionists are attempting to set back the interests of the slave catchers and thereby the interests of the slave owner, but these interests are not rightfully held, and so they do not do wrong by setting them back—indeed, what they do appears praiseworthy. By preventing the abolitionists from causing these permissible harms, I am defending interests that are not rightfully held. Surely *that* is morally wrong. The point here is that if harming is not always wrong, then prevention and defence against harm is not always permissible.

Coming back to war, defence and defensive weapons, an aggressive state A that invades another B in order to steal its resources, occupy its land and subjugate its people, and thereby wages war, will normally not, in the conduct of that war, remain on the offensive the whole time, and certainly will not if the war is of any significant duration. B will fight back and A will need to defend its own assets. If B fights back by trying to kill civilians, then A's defence of its citizens is permissible, but if B fights back by attacking A's armed forces, then A's defence of the those forces is not permissible. A's armed forces are no longer immune from attack, once they have been used for the ends of aggression. In a war of any significant duration in a large theatre of operations, there will be periods when A is on the defensive with B attacking. This was the case in France, North Africa, Italy, the Western Pacific, Burma and of course the Soviet Union in World War Two, and from the middle of 1943, Germany defended almost continuously, as did Japan from 1944. A useful way to describe the ebb and flow of such conflicts is to use the language of the levels of strategy. Here I quote from Edward Luttwak[8]

> [At the first level there is the] *technical* interplay of specific weapons and counter weapons [that is] subordinated to the *tactical* combat of the forces that employ those particular weapons...the tactical level moves of particular units of armed forces on each side are merely subordinated parts of larger actions involving many units, and this is the *operational* level ...Events at the operational level can be very large in scale, but never autonomous. They are governed by the broader interaction of armed forces as a whole within the entire theatre of warfare...[the] higher level of *theater strategy*...The entire conduct of warfare and peacetime preparation for war are in turn subordinate expressions of national struggles that unfold at the highest level of *grand strategy* (Luttwak 1987: 69–70).

With the exception of the first, *every* level of strategy can be defensive or offensive, depending on the aims, objectives and circumstances of the state in question.

The technical level is concerned with the nature and performance of individual weapons and how they (would) fare against other available weapons. For example, how fast an airplane can fly, how far, how manoeuvrable it is, what weapons it carries and how accurate they are are technical considerations, as are how it matches up against other airplanes, anti-aircraft weapons, and so on. If the hypothesis suggested in the previous section is correct, then whether the airplane is a defensive weapon will depend on the context in which it finds itself and what else is available. The German Messerschmitt 109 fighter planes of the Second World War fought against the British Spitfires and Hurricanes and were fairly evenly matched. These engagements were offensive until 1942, in the Battle for Britain, the Battle for Malta and in North Africa, then, as the tide of the war turned, their role was primarily defensive: the same airplanes fighting one another in similar ways, but in different contexts. The ways these adversaries fought one another is thus part of tactics, of particular units engaging one another in various ways. For instance, the Spitfire pilots learned tactics when fighting against Messerschmitts, but when they were transferred to the Pacific to fight Japanese Zero fighters, they found that their methods did not work because the Zeros were much more manoeuvrable, and they had to adopt new tactics (Smith 2015: 146–149). More illustrations could be given about the way tactics are incorporated into operations, and how operations whose objectives are offensive may need defensive tactics from time to time, and how such operations can be used to enact defensive theatre strategy, and how defence and offence are thus interdependent. But I now want to move up to the level of grand strategy.

Luttwak says that "national struggles that unfold at the highest level" are the province of grand strategy. Concerns that states have at the 'highest level' have to do

---

[8]As I have done before, see Forge (2012, Sect. 8.4) for more discussion of this passage.

with security, among others. The *raison d'être* of the state is to defend the people, who in turn allow the state a monopoly on power.[9] Another authoritative source, Martin Wight, expresses this in terms of 'vital interests' and tells us that "There are certain things that a power [a state] deems essential to its continued independence; these are its vital interests, which it will go to war to defend" (Wight 1979: 95). Grand strategy thus encompasses the overall ways in which states go about ensuring that their vital interests are protected and promoted. Wight also tells us that a state's vital interests are what it takes them to be, and in this sense they are a matter of the state's perceptions rather than the result of any objective measure, which is not to say that it is not possible to judge that states make mistakes about such things. Wight, and others who can be regarded as Realists when it comes to international relations, do not entertain concepts such as those of just war. They do not consider aggression to be just or unjust, only whether it succeeds or fails to defend vital interests. But we are not adopting this way of looking at things, for the moment at least, and so we judge grand strategies by the standard of JWT and only allow the resort to war to be just if it is self or other defence, and this *is* to use an objective measure.

As examples of contrasting grand strategies which both gave rise to aggressive wars, compare those of Germany in years up to 1914 with those of Germany from 1933 to 1939.[10] In both instances, Germany invaded France and the Russia/Soviet Union and initiated a world war. However, in the first instance, Germany believed that instability in the Balkans caused by the assassination of Franz-Josef would lead to conflict between its (weaker) ally Austria-Hungry and a member of the opposing alliance Russia, which mobilised its troops on the same day as the assassination. Germany believed that this would lead to a wider European war and that it would be at a great disadvantage if it had to fight on two fronts: against France and Britain, the other two members of the opposing alliance, in the west, and Russia in the east. War plans developed by the German general staff called for Germany to attack and defeat France first, beginning with a invasion of neutral Belgium, and then turn on Russia. This was a disaster, with Germany eventually losing a war of attrition: no one knows what would have happened if Germany had not started the war. German grand strategy developed by Hitler after the Nazis came to power was aimed at securing resources to ensure that Germany would never again be defeated because it ran out of materials and supplies—this was the accepted view as to why the country had lost World War One. Until 1939, Hitler was able to realise his aims—

---

[9]In Chap. 2 I introduced Posen's idea of military doctrine as the element of grand strategy that has to do with security, security to be enforced by military means. This characterisation is entirely consistent with Luttwack's, given that military doctrine is taken to encompass the levels of strategy below that of grand strategy, in this sense that military doctrine determines, or should determine, their substance. More on this in Chap. 6.

[10]There is an immense volume of scholarship about the origins and causes of both world wars, with new and important analyses becoming available. I make no claim to be familiar with even a fraction of this literature. But my aim in this paragraph is simply to give an example of how quite different, even opposite, basic motivations and aims can give rise to essentially the same actions, namely aggressive war.

re-militarise the Rhineland, forbidden under the Treaty of Versailles in 1919, achieve union with Austria and annex the Sudetenland and dismember Czechoslovakia—without provoking France and Britain into war. However, the German invasion of Poland did lead to another European war. Unlike World War One, Hitler's pretence that he was protecting ethnic Germans living in Poland was a smoke screen and his was a grand strategy of aggression and conquest, with the aim of substantially increasing the territory of Germany.

If a weapon is used for a defensive mission, to protect an asset, it does not follow that this use is permissible. It is only permissible if the mission is conducted in the course of a just war. On the other hand, if a weapon is used on an offensive mission, to attack an asset, it does not follow that this use is impermissible. It may be permissible if the mission takes place in the prosecution of a just war. It follows that if there were purely defensive weapons, then any missions they were employed in —which must by definition be defensive—would not, for that reason, be permissible; that is, they would not be permissible *because* they were defensive. For our purposes, the existence or lack thereof of purely defensive weapons is therefore not decisive: even if there were purely defensive weapons, weapons research that aims to design such weapons would not be morally permissible for that reason. This is because, as we have just seen, defence is not always morally permissible. We can now state with confidence that there are no kinds of weapons research which are justifiable, and hence permissible, *independent* of the contexts in which the weapons are to be used. Put another way, and referring back to Luttwak, we can say that the justification of weapons research must move beyond the technical level. And, finally, this discussion of defence vindicates our determination that the *only* primary purpose of a weapon is that it is the means to harm.

4.5 Deterrence

Deterrence was briefly discussed in the previous chapter, where we saw that it is a derivative function of a weapon. In this chapter, we have established that weapons research can only be justified with reference to the actual uses that the weapons created have or will have, and that there is no justification possible simply in terms of technical characteristics, for instance, that the weapon in question only has a defensive role (were that even possible). So if weapons research is to be justified, then it will be necessary to appeal to actual defensive or deterrent functions that the weapons produced will have. At the beginning of the next chapter, I will spell out a litmus test or standard by which such justifications are to be evaluated: if they fail to live up to the standard, then they fail as justifications and the judgement that the weapons research is morally wrong is not revoked and the research remains morally impermissible. To understand what these amount to, we need to know what a defensive or deterrent role is. Enough has been said about defence, so now we need to say more about deterrence. I begin by reviewing the idea of deterrence, and then I will give an

extremely brief account of nuclear deterrence in the Cold War to illustrate four
problems with conducting weapons research for the purposes of deterrence.

Recall that deterrence is a relationship between two or more states—as before I
will begin by considering the simplest arrangement where there are just two states
involved, A and B—such that B believes A intends to undertake some venture that
is against B's interests. B wants to prevent A from acting and so institutes measures
that it believes will deter A: B believes that A will calculate that the benefits of
acting will not be worth the costs B will impose, and so will be deterred from
acting.[11] Not every instance of deterrence will entail implied threats to use force,
but those will not be of interest here since we are talking about weapons and threats
of violence. The costs imposed by B will be the harms that occur when weapons are
used, costs of war and conflict. Thus B may threaten war or imply that it will go to
war with A if A carries out its intended course of action, or what B believes A
intends to do. This will typically come about, as Wight has informed us, when B
believes its vital interests are at stake.[12] It will not normally, or ever, happen that A
just wants to wage war for the sake of it; to get what it wants it may need to go to
war, or this is what B believes, but war will not be an end in itself.[13] In order to
deter A, B may believe that it must do whatever is necessary to make the costs of
going to war too high for A, for instance, build up formidable armed forces. We can
therefore distinguish between what B aims to do, or rather prevent, and the means
or methods by which it seeks to achieve this end. What happened during the Cold
War was that aims and methods became conflated, as we shall now see. This is a
most salutary lesson for those who would acquire weapons for the ends of
deterrence.

Nuclear weapons became a means for deterrence in the mid-1950s for the US.
This was made explicit in a US nuclear doctrine known as massive retaliation.[14]
Lawrence Freedman tells us that "massive retaliation was widely assumed to be
founded on an undiscriminating threat to respond to any communist-inspired
aggression, however marginal the confrontation, by means of a massive nuclear
strike against the centres of the Soviet Union and China" (Freedman 1989: 76). At

---

[11]The opposite of deterrence is compellence, when B seeks to coerce A in doing, rather than not
doing, something. We will not be concerned with compellence here.

[12]If vital interests are those B is prepared to go to war over, and B is prepared to go to war only if
its vital interests are at stake, then perhaps these ideas are in need of further clarification if they are
to be truly explanatory. Carl von Clausewitz' famous dictum on war is that it is politics/policy
carried on by other means, namely violent ones (von Clausewitz 1984: 7). If we understand the
states highest political concerns to revolve around its vital interests, then when peaceful ways of
securing these fail, states resort to 'other means'.

[13]Warlords like Napoleon create exceptional circumstances.

[14]Nuclear doctrine is a special form of military doctrine, and it comprises authoritative statements
about the role of nuclear weapons in war. Since nuclear weapons are so very dangerous, the
superpowers, and other nuclear armed states, realised that they needed to spell out how they might
be used, to try to forestall any misunderstandings. The nuclear strategy of a state would then
amount to an overall plan about how to enact nuclear doctrine in various contingencies, see
Freedman 1989, *passim*.

that time, the US had a clear superiority in nuclear weapons technology and so felt that it could use this to threaten the Soviet Union (USSR) and China with nuclear attack if either state were to act aggressively outside their 'spheres of influence'. The prevailing view in the US was that the USSR and China were inherently aggressive states because they believed that communist revolutions would come about in capitalist states, and were not above helping the revolution along a bit if they could.[15] US grand strategy in the mid-1950s, which relied on nuclear weapons, was primarily focussed on deterring communist aggression, on 'containing' communism: containment was the end of deterrence and the threat of nuclear attack its means. The USSR and China, and everyone else, knew what nuclear weapons could do, so the threat to use large numbers of nuclear weapons to attack centres, which included cities, was unprecedented and terrifying. The US had previously attacked Japanese cities with all the atomic bombs it had, namely two, but now it had a lot more and these were huge thermonuclear weapons.

Stalin was aware as early as 1946 that the USSR had to match the US nuclear capability, and by the mid-1960s it had managed to do so, managed to achieve 'parity'. This meant that massive retaliation was obsolete because now the USSR could retaliate back. Both superpowers had nuclear-armed ballistic missiles, against which there was, and is, no defence. In other words, the aim or ends of massive retaliation could no longer be achieved by the means. Moreover, both superpowers by now had so many nuclear weapons that some at least would be able to survive a surprise 'counterforce attack', a nuclear strike aimed at the opponent's missiles, and so it seemed that each side had 'survivable' nuclear forces. The question was: what to do with them! The answer, really the only answer, was *nuclear deterrence*. I define nuclear deterrence to be a deterrent regime in which the means and the ends are nuclear weapons: the means and ends are conflated. That is to say, the aim of nuclear deterrence is the prevention of nuclear attack and the means are the threat of nuclear attack. Given that both superpowers subscribed to this notion, then the result is *mutual* nuclear deterrence. This was not the only proposal about what to do with nuclear weapons, with some believing that they could actually be used to fight and win wars, but that never became official nuclear doctrine. As a consequence, there continued to be a build-up of nuclear forces for the next fifteen years in order to ensure that enough would survive as a retaliatory force against any conceivable surprise counterforce attack. And this led to nuclear arsenals that could destroy the whole world. The Cold War is over and the USSR no longer exists, and while there are considerably fewer nuclear weapons now than there were thirty years ago, there are still enough to kill almost everyone on the planet and destroy everything of value.

The history of nuclear deterrence reveals the central problem, and the first of four problems, with undertaking weapons research in order to provide the means for

---

[15]There is a vast literature on this topic. What I write here is a very simple sketch, for the purpose of illustrating the problems and paradoxes that arise when nuclear weapons are used for deterrence. I have written more on the topic elsewhere, see for instance Forge (2012: 93–98).

deterrence. It is that deterrent regimes come and go but the means in the form of weapons remain. In the case of nuclear weapons, these became the central focus of deterrence, having been introduced as a means to ensure deterrence—the solution became the problem. It is true that there are not nearly as many nuclear weapons as there were, but more states now have nuclear weapons and Russia, the US, China and other countries could quickly produce more if they felt the need to do so. Knowledge of how to produce weapons in the form of designs remains, once weapons research is successful, even though the weapons themselves may no longer be operational. Nuclear weapons are the paradigm case because they are so very dangerous. As I have said, it is generally agreed that nuclear weapons are not the means to fight wars because they are so destructive. There is no rational purpose to destroying cities, killing all the inhabitants and leaving them as unapproachable radioactive ruins.[16] The only role for nuclear weapons remains nuclear deterrence, even though the states that possess them are not in any direct conflict with one another, and have much more to gain by fostering trade and other forms of cooperation than in conflict. It is the weapons that are the problem. And this has come about by mistakes, misperceptions, stupidity, and driven by the standard rationale for weapons research.[17]

We saw in Chap. 2 that nuclear weapons were originally promoted by a small group of European scientists who had first-hand experience of German physics and German politics in the 1930s, and who had fled to other countries, notably Britain and the US, to escape from persecution. They thought that the new field of nuclear physics might become the basis of a new kind of weapon and sought to inform the governments of Britain and the US of this, in case German scientists had the same idea and that the weapons become available to the Nazis. The very rationale behind the first proposals for looking into the possibility of nuclear weapons was deterrence. And deterrence has remained the *only* role for nuclear weapons, save for their use on Japan in 1945 which remains one of the greatest war crimes of all time. The irony here is that a German nuclear weapon was never possible because of the extreme difficulty of obtaining the fissile material and of designing the first bombs. Not only this, when a small team of German scientists examined the question and did some experiments, they made a crucial mistake and believed an atomic bomb could not be made. And this is another, more general problem, with deterrence: how can one be sure that deterrence is required, why does B believe A wants to do things that will compromise its vital interests? There are all sorts of possibilities here, and lots of room for mistakes. The balance of opinion now is that the USSR never intended to extend its influence by aggressive means but rather set up the regimes in

---

[16]Unfortunately certain groups such as ISIS would welcome such destruction, as the 'end of days'.

[17]See Lebow and Stein (1994) for a catalogue of these mistakes.

Eastern Europe as a defensive buffer against western encroachment, given its experiences in the Second World War.[18]

This brings us to a third problem with deterrence. If A has no intention of trying to coerce B with a view to advancing its interests and sees B building up its armed forces and claiming that this is to deter A, then A has every right to become suspicious of B's motives. So A in turn may feel it needs to build up its armaments in return. And again, this happened in the Cold War, with the USSR matching US nuclear developments. Robert Jervis coined the term "security dilemma" to refer to a situation in which state A contemplates the weapons developments of state B and wonders whether to match those developments, knowing that this might provoke B into another new round of weapons acquisitions (Jervis 1978: 187). Such tit for tat arms build-ups then makes the security situation more perilous for states because they are both equipped more advanced weapons. Misperceptions about deterrence can trigger such a state of affairs, and lead to an arms race. Now to the last problem about weapons research and deterrence that is illustrated by nuclear deterrence in the Cold War, and that is the focus on capabilities and worst case scenarios. The worst case scenario for either the USSR or the USA was a surprise nuclear attack aimed at destroying the nuclear deterrent—a so-called counterforce strike. If successful, such an attack would leave the recipient without the means to retaliate. This is 'worst case thinking', but it is not what any sensible or rational person outside the strange world of nuclear strategy would ever contemplate in reality: why would either side ever want to destroy the other completely?[19] However, the *possibility* of a successful strike was deemed destabilising—i.e. it might encourage such an attack —so in the interests of *stability*, it was thought that only a vast array of all different kinds of nuclear weapons hidden in all sorts of places was called for. Ensuring nuclear deterrence thus drove the arms race. In the light of all these issues, it is clear that deterrence is not a good reason to undertake weapons research, and hence no basis at all on which to argue that weapons research is justified.

4.6 Conclusion

We have agreed that the only justification for causing harm to those who have a right not be harmed is the prevention of further harm. This does not license unlimited harming: it is not acceptable to cause a very great deal of harm to forestall a much smaller amount of harm, though just what the limits are presumably varies from case to case. We have also agreed that it is wrong to provide the means to harm, we have agreed to accept the means principle, MP. At the end of Chap. 2,

---

[18]And as we have seen, the same weapons and the same forces can be used for defence and attack; so no inferences about aggressive intent can be make on the basis of the composition of military forces.

[19]The collateral damage would cause huge civilian loss of life and the resulting fallout would cause problems for the attacker as well, possibly a nuclear winter.

where this principle was introduced, I considered an objection that weapons research aimed at producing weapons for defence did not conform to MP and this suggested that the claim that weapons were the means to harm, that being their primary purpose, was not correct. It was not correct, according to the objection, because there are weapons that prevent harm, and so whatever we make of MP, it does not cover all of weapons research. In this chapter we have seen that this objection cannot be sustained. It cannot be sustained, not because weapons cannot be used to prevent harm, but because this is not their primary purpose. And we have seen that we cannot simply focus on the technical capabilities of particular weapons and draw conclusions about whether or not their deployments accord with the kinds of moral constraints one would like to impose on wars and conflicts. We have seen that defensive missions and deployments can aid aggressive unjust wars. In sum, one cannot draw any conclusions about whether weapons prevent harms that ought not to be caused without looking at the actual historical circumstances in question. There is, to put the matter another way, no *ahistorical justification* for weapons research.

Appendix: ABM Systems

An Anti-Ballistic Missile system is a weapon to be used to shoot down or otherwise intercept ballistic missiles. There have been ballistic missiles that are not armed with nuclear warheads, such as the German V2 and various conventional Scud missiles that were used in the First Gulf War and others besides, but the main concern with ABM systems was (and is) with nuclear-armed ballistic missiles. A ballistic missile is aimed and fired and then proceeds to its target first of all under the influence of burning rocket fuel used to accelerate it and then under the influence of its inertia and gravity. The weapon is not guided in that there are no systems that identify and lock onto a target and then direct the path of the missile after it has been fired, and ballistic missiles share this feature with all other projectile weapons.[20] In theory one could have ABM interceptors that were also ballistic. But this would require the most precise calculation of the flight path of the incoming warhead and then the most precise determination of the flight path of the interceptor, and this would need to take account of factors that in practice could not be determined, such as wind speed. In practice, the interceptors needed to be guided to their targets. This is also extremely challenging: the interceptors have to be very

---

[20]That is not quite true for every kind of ballistic missile. So-called MIRVed ballistic missiles carry more than one warhead with are independently targetable. These are carried on a 'bus' which has some manoeuvrability: the warheads can be released with slightly different trajectories and hit targets some hundreds of kilometres apart. There were a few MARVed missiles, such as the Pershing 11. These had manoeuvrable warheads, so they did actually seek out their targets, but none are now in service.

fast and highly manoeuvrable and they have to be fired at the right time and in the right direction.

Dietrich Schroeer tells us that an ABM system has to fulfil five functions

1. The enemy warhead has to be detected; 2. the re-entry vehicle carrying the warhead has be to distinguished from the missile debris and penetrations aids such as decoys and chaff; 3. the path of the warhead has to be predicted into the future; 4. the interceptor has to be guided to its target; and 5. the incoming warhead has to be destroyed in a verifiable way by the explosion of a nuclear warhead (Schroeer 1984: 238).

Attacking nuclear warheads are therefore to be destroyed by defending nuclear warheads, the idea being that only the latter were able to definitely destroy the incoming weapons. Defending in this way demands a substantial price, in terms of nuclear fallout and other damage, such as massive disruption to communication, but presumably better that actually receiving the incoming warheads. The US began to develop a two-layered system based on these principles in 1959, which was partially deployed in 1971, but abandoned a year later under the 1972 treaty to ban ABM systems. A two-layered system is one that comprises two completely independent subsystems sharing no common components and operating at different ranges, the idea being that the warheads missed by the longer range interceptors would be mopped up by the shorter range weapons.

I said I would consider ABM systems as a possible example of a purely defensive system, having claimed that what counts as a defensive weapon depends on the context in which it is in place, from which it follows that there are no purely defensive systems. Having made the point that defensive missions, tactics, etc., are (usually) necessary for aggressive wars of conquest, it may seem that we do not really need to worry about pure defence anymore, because that is now seen not to be able to justify weapons research. However, this example is instructive because of the way in which it confirms the view that what counts as a defence depends on the context, and in the case of ABM systems, the context is informed by strategic or even grand strategic considerations. I will get on to that issue in a moment. Note first that the technologies needed to set up an ABM systems have much in common with those needed for an offensive nuclear missile capability. These include the nuclear warheads for the interceptors, which must be small, be able to detonated at just the right moment and resilient enough to stand up to the high acceleration of the interceptor missiles. The missiles themselves were, for the US system just mentioned, adapted from submarine-based missile technology (Schroeer 1984: 237). Thus the interceptors could be turned back into offensive systems with some modification.

The reason why it was possible for the superpowers to agree to a ban on ABM systems was that they were generally accepted to be extremely *destabilising*, that is to say, make nuclear war more likely. The reason ABM systems are destabilising is because they would not be 'leak-proof', that is, they could never in practice be guaranteed to intercept all incoming warheads: some were bound to get through.[21]

---

[21]This is why I have said that there is no defence against ballistic missiles.

By contrast, a leak-proof system would be profoundly stabilising because nuclear weapons would have been rendered, to use the famous words of President Ronald Reagan, impotent and obsolete (providing everyone had such a system). But an imperfect 'leaky' system was thought to encourage the possessor to launch a first strike at its opponent's nuclear arsenal, either the US or USSR given that we are talking about the period of the Cold War. Assuming that such a strike is not completely successful and retaliation follows, the thinking was that the ABM system would manage to intercept most of the surviving warheads, and damage caused by the relatively few that got through would be 'acceptable'. The advantage would be that an implacable and dangerous enemy, either the US or USSR, would have been eliminated. This way of thinking is surely very strange: any form of reckoning that countenances just one nuclear warhead landing on one's territory—on Moscow or New York for example—has gone off the rails in a serious way. But such was Cold War nuclear strategy, and it demonstrates clearly how defence against nuclear weapons could not be separated from the management of offensive systems: the context is everything.

References

Kenyon, K. 1960. *Excavations at Jericho*. Jerusalem: British School of Archaeology.
Forge, J. 2012. *Designed to Kill: The Case Against Weapons Research*. Dordrecht: Springer.
Freedman, L. 1989. *The Evolution of Nuclear Strategy*, 3rd ed. London: McMillan.
Jervis, R. 1978. Cooperation under the Security Dilemma. World Politics. 30(2), 167–214
Lebow, R., and J. Stein. 1994. *We All Lost the Cold War*. Princeton: Princeton University Press.
Lee, W. 2016. *Waging War*. Oxford: Oxford University Press.
Luttwak, E. 1987. *Strategy: The Logic of War and Peace*. Cambridge, Mass.: Harvard University Press.
Smith, P. 2015. *Mitsubishi Zero: Japan's Legendary Fighter*. Barnsley, South Yorkshire, UK: Pen & Sword Books.
Stolfi, R. 2014. *The 7th Panzer Division in France and Russia*. Barnsley, South Yorkshire, UK: Pen & Sword Books.
Schroeer, D. 1984. *Science, Technology and the Nuclear Arms Race*. New York: Wiley.
van Creveld. 1991. *Technology and War*. New York: Free Press.
Von Clausewitz, C. 1984. *On War*. Indexed edition. Edited and translated by M. Howard and P. Paret. Princeton: Princeton University Press.
Wight, M. 1979. *Power Politics*. Harmondsworth: Pelican

Chapter 5
No Justification for Weapons Research

If there can be no ahistorical justification for weapons research, then all attempts at justification must be *historical*: they must refer to the situations and circumstances in which weapons are actually used. In view of the general condition for all such justification, in terms of harm prevention, it must be the case that the harms caused by the use of the weapons produced must also prevent harm, and moreover, the harms caused must not be excessive and the harms prevented must at least be comparable to, and preferably much less than, those caused. But the problem here is that weapons designers *cannot know* what is going to happen in the future when it comes to the ways the weapons they design are used (if indeed they are ever used). If they cannot know what they *must* know in order for their work to be justified, then justification of weapons research will not in fact be possible. I am going to spend this chapter and the next making good this claim. If I am able to do so, then I will have shown that weapons research is never morally permissible: that is to say, in *no circumstances* and under *no conditions* is it ever morally acceptable to undertake weapons research. The case against weapons research will have been made.

I will begin this chapter by asking how one might make a judgement to the effect that causing some harm can prevent greater or at least comparable harms. This is the principle, but how does it work in practice? It is not straightforward even in what we might call paradigm cases, which involve well-understood and carefully arranged situations. When it comes to weapons research, it would appear that the problems are magnified to a degree where the only conclusion that we could endorse is that such a judgement can never be made on any remotely plausible grounds. I will explain why this is. But before doing so I will raise the following possibility: suppose one considers particular *kinds* of contexts in which weapons research is to be conducted, for instance, under the auspices of democratic states. The suggestion here is that democratic states do not embark on aggressive wars and hence not only are the wars they do engage in justified, so are the things that they need to do in order to defend themselves (and others) from aggression, and this includes weapons research. In other words I will assume that what lies behind this way of thinking is that democratic states only fight just wars, that if fighting a just

© The Author(s), under exclusive licence to Springer Nature Switzerland AG 2019
J. Forge, *The Morality of Weapons Research*, SpringerBriefs in Ethics,
https://doi.org/10.1007/978-3-030-16860-5_5

war is permissible, so is doing what it necessary to fight a just war (cf Forge 2012: 191). It turns out that the problems with this way of thinking about weapons research, and the attempt to restrict the context of the activity, are of precisely the same kind as those that face the weapons researcher in any situation whatever, regardless of the nature of the state in which she finds herself in. To address these issues we will need to return to JWT.

5.1 Preventing Harm

There are paradigm instances of justified harming which involve medical or dental procedures and these represent a good starting point for our discussion (for example, Gert 2005: 224). Thus, a medical or dental procedure should be carried out with the consent and in the interest of the patient, with every effort made to reduce the amount of harm—which we can identify here as pain—in order to reduce or prevent greater pain and suffering in the future. The procedure is justified if these conditions are satisfied. Mistakes can be made: the procedure might not have been necessary after all, the patient not fully informed, it might not have been carried out correctly and the pain caused not managed properly. For instance, if a patient has a serious and life-threatening illness, such as metastatic cancer, the decision to have any treatment at all, what treatment to have, when to stop or continue with it, can be extremely difficult.[1] Sometimes the treatment can be worse than the disease and the patient is worse off than she would have been otherwise. When this happens, it depends on the circumstances whether the harm caused was not justified and those involved blameworthy. Such matters can be complicated and will turn on whether anyone was at fault due to negligence, careless or incompetence. Such examples illustrate three things: first of all, honest mistakes can be made when attempting to prevent or reduce future pain, secondly, mistakes can be made which are not honest in the sense that the agent(s) are blameworthy, and finally and most importantly for present purposes undertaking actions to prevent or reduce future pain or harm is by no means a simple and straightforward matter even in routinely-undertaken and well-understood circumstances.

If someone needs to decide whether to embark on such a course of action where she causes harm, she needs good reasons *beforehand* to believe that she will suc-ceed, namely prevent or reduce future harm. In view of MP, we can also impose this requirement on any attempt to justify weapons research, namely that the relevant assessments must be made *before* the agent begins her work, or at least at a time when she can stop before it is too late, before she has done enough to make a contribution to a viable design. A necessary part of having good reasons here is to know *how much* harm the agent is going to cause, how much she needs to cause, for if *that* is not known, then no comparison can be made between the harms caused

---

[1]For more details and many examples, see Brody (1988).

and prevented. And there is another matter that needs to be dealt with at the outset, and that is how to determine or *calculate* the harms caused, to determine their 'amount', for if the aim in causing harm is to prevent, reduce or avoid future, greater harm, then it must be possible to compare the former to the latter and judge that on balance more harm has been prevented than caused. We have agreed that it is unacceptable to cause more harm than is prevented, reduced or avoided. It is worth making this quite explicit in principle HC (Harm Caused):

> HC: An agent must be able be make a reliable estimate of the amount of harm she is going to cause before undertaking a justifiable action that is intended to prevent or reduce future harms.

One assumes that medical practitioners, and other specialists, have considerable experience to draw on when advising their patients of the probable and possible side-effects and risks of operations and other treatments. So in the 'paradigm cases' we would hope that HC can be satisfied. And this is just the first step. It is also necessary to estimate the harms prevented and then a comparison between these and the harms caused must be made—I will not discuss this further, as it is clear in outline at least what these estimates will be like.[2] When it comes to weapons research, it is not even possible to move beyond the first stage and show that HC is satisfied.

Let us ask just what needs to be known, or estimated or even guessed at, in order to work out how much harm is going to be caused by a given weapons research project. The harm is going to be caused by the weapons which the project succeeds in designing, assuming that the project is successful. Weapons cause harm principally when they are used in wars and other conflicts, when they are used to enact violence. It will therefore be necessary to know, or estimate, who will use the weapons, how they will use them, and where. It will be therefore necessary to predict which wars and conflicts they will be used in and how they will be used. Given this demand, it is already clear that *no one* can have all the relevant knowledge or information, let alone a weapons researcher who is presumably not expert in forecasting. The outbreak of war and the conduct of war are among the least predictable human activities, and as a consequence the task of determining the way a particular kind of weapon will be used in future wars will be impossible. If this information is required before weapons research can be judged to be morally permissible and if it is always impossible to get it, then it follows that weapons research is never permissible. End of story!

I believe that the 'story' really does end here, but it is still necessary to complete it by giving some more examples from the history of weapons development which confirm my claim about lack of predictability of uses. But there is another approach

---

[2]But is often not clear how to decide what to do even when it seems that the harm to be caused and the harms prevented are relatively clear. For instance, if medical treatment promises to prolong a someone's life at the expense of the quality of that life, different people may make different judgements about how these prospective outcomes are ranked.

which I have foreshadowed, and that is to restrict the context of weapons research. The idea is this: suppose there were some states that only fought just wars and only undertook weapons research in order to equip themselves to fight just wars, to resist aggression. Would not weapons research done in such contexts be justified? This would be to concede, tacitly, that not all weapons research is justifiable, but at least some is. This idea resonates with the standard rationale for weapons research in terms of defence, but now it picks out certain kinds of states, 'good citizens of the world', who always treat other states fairly and not aggressively. But this suggestion founders not simply because there are no such states, no states that only fight just wars, but because even if there were, weapons research is not consistent with JWT, as we shall now see.

5.2   Just War Theory and Democracies

JWT underwent a revival—it is the modern version of an ancient tradition—with the publication of Michael Walzer's *Just and Unjust Wars* in 1977. I will therefore concentrate on those parts of the theory that are relevant for the present purpose and refer the reader unfamiliar with it to some of the large body of recent work on the topic.[3] To begin, it is clear that at most only one side (state or coalition of states) can have just cause for war. It is clear because by definition the only just cause for war is resisting aggression, and if B is resisting aggression by fighting A, then A must be the aggressor. Neither side could have just cause if both are aggressors. If B has just cause for war, then it does not follow, assuming B goes to war, that it is fighting a just war. Having just cause is not sufficient here, because there are other *ad bellum* conditions, conditions that determine whether the resort to war is justi-fied, beside just cause, and there are *in bello* constraints, that is, constraints on how the war is fought. There are therefore a number of permutations and combinations: for instance, state that is justified in embarking on war may not be fighting a just war because it violates an *in bello* constraint, and conversely.

A state that starts out being engaged in a just war, because it satisfies all the conditions and constraints of JWT, may end up violating some of them, which means that the war is not longer just. This is what happened in World War Two to Germany's three main protagonists, the USSR and Britain, and the USA. Germany attacked the former two states in 1940 and 1941 respectively. By the end of 1940, Britain was deliberately targeting German civilians by strategic bombing, namely,

---

[3]Brian Orend's *The Morality of War*, Orend (2006), is a clear exposition of JWT which follows Walzer's original formulation. Larry May (ed.) *The Cambridge Handbook of The Just War*, May 2018, is the most recent (at the time of writing) of a number of books published since Walzer (1977). In the foreword and in the introduction to this collection it is mentioned that there has been renewed interest in JWT in the present century and in particular so-called revisionist Just War Theorists have challenged some elements to the orthodoxy, for instance in regard to the moral equality of soldiers.

indiscriminatingly bombing German cities. This violates the *in bello* constraint which forbids deliberately attacking civilians.[4] Millions of Red Army prisoners of the Germans died of neglect and mistreatment in that war. This violates the *in bello* constraint which requires the protection of enemy soldiers who have been captured or surrendered: they are no longer combatants and should be given 'quarter'. The USSR reciprocated, though not quite to the same degree—they did not kill quite as many German prisoners. When the Red Army reached Germany, it indiscriminately killed and raped German civilians, as had the Wehrmacht in the Soviet Union. Turning to the USA, it participated in the strategic bombing of Germany from 1942 and of course committed the two greatest single violations of JWT in history when they dropped atomic bombs on Hiroshima and Nagasaki.

I mention these examples for two reasons. The first is that a state may start out in war by 'playing by the rules', by abiding by the principles and constraints that are intended to minimise the worse excesses of war and confine violence to those whose business it is to fight, but end up doing whatever it takes to win. I will refer to the latter as *military necessity*.[5] Hitler, for instance, made it clear from the beginning that military necessity was the only guiding principle in his war on the USSR: his Commissar Order, which stated that the political officers, the commissars attached to all Red Army units, were to be summarily executed, was in accordance with military necessity as he understood it. The atomic bombing of Hiroshima and Nagasaki was also based on military necessity, as understood by Truman: he wanted to force Japan to surrender unconditionally by whatever means available. Military necessity and JWT are thus incompatible. My second reason is that democratic states, states that one might expect abide by the laws of war, do not always do so.[6] Moreover, democratic states not only resort to military necessity when engaged in desperate struggles against wicked enemies, they sometimes wage war without just cause from the very beginning. This was the case, for instance, for the US in Vietnam, and for the US, Britain and Australia in the second gulf war.[7] We can therefore conclude that weapons research cannot be justified with reference to contexts restricted to those under the control of democratic states, even if it were

---

[4]A distinction is now made between non-combatants, people who do not fight but nevertheless contribute to the war effort, for instance by working in armaments factories, and civilians who make no such direct contribution. The former are (sometimes) considered to be 'legitimate' targets. The British, Churchill in particular, claimed that bombing German cities was the only way to fight back and moreover it was an acceptable tactic because it would weaken the German people's will to continue the war. It didn't and it was no excuse.

[5]I should stress that this is *not* the standard sense of the term, which is usually taken to mean that any measures can be used in war that are necessary for achieving objectives, provided that they do no violate international humanitarian law and that the objectives are legitimate. My usage drops the latter two qualifications.

[6]And I assume one would think them more likely to do so than other kind of states.

[7]Parliamentary democracy in England was certainly in place by 1832 with the Reform Act, but there were parliaments in England from 1265. England, and then Britain, waged many unjust wars of colonisation and conquest, right up until the First World War. Many more examples of democracies doing the same can be given.

possible to quarantine the outcomes to use by those states alone. Democratic states are no better than anyone else when it comes to playing by the rules.

If actual instances of just wars are hard to come by, and I believe they are (Forge 2012: 221), then one may wonder about the status of JWT: is it intended to provide moral guidance for those who have the power to initiate conflicts, or is it a means for making ex post facto judgements about the morality of wars, by philosophers or historians perhaps?[8] I am inclined to the latter position because I believe that relationships that actually exist between states in the world are best described by the Realist account of international relations. However, I also think that if wars are to be just or moral or ethical, whatever word one wants to use here, then JWT, or something very similar to it, provides the correct description of what such wars must be like (more on this in the appendix to this chapter). Thus I think it is right to demand that the resort to war is only just if it resists aggression, that it must be conducted in such a way as to spare civilians as much as possible, to treat prisoners fairly, and so on. The fact that this does not happen very often simply means that there are not many just wars, not that JWT is false. But if wars are rarely just, then this makes the strategy for justifying weapons research by restricting it to contexts governed by states that only engage in just wars tenuous indeed. How can the weapons researcher decide whether the weapons she creates will be only used in just wars, assuming again that it is only the armed forces of her country that will have access to them, if she has no reason to believe that her country only fights just wars? I don't think she can do this, but even if she could, it would be to no avail because weapons research is not compatible with JWT, as we shall now see.

5.3   Just War Theory, Proportionality and Weapons Research

In the introduction to this chapter I said that if a state is justified in waging war, then it seems it is justified in doing what it takes to wage war, such as recruiting and arming its soldiers. Developing weapons, doing weapons research, would seem to be a necessary part of that process. We have also seen that JWT puts limits on what can be done when it comes to fighting wars: JWT curbs 'military necessity', for example. It is therefore not true that when a state has just cause, thereafter anything goes. It is also not true that having just cause is sufficient reason for going to war; there are a number of other *ad bellum* conditions that need to be satisfied before a state is justified in going to war. One of these is proportionality, which is going to

---

[8]This topic is raised in my detailed discussion of JWT in Forge (2012): see Chaps. 10, 11 and 13 of that book. In his foreword to May 2018, Jeff McMahan, having noted the changing character of warfare in the last two decades, suggests that much recent writing on JWT has been internal to the field, with philosophers addressing other philosophers, and not with the developments he mentions. This does not surprise me. And I note that this most recent collection on JWT has nothing at all to say about *ad bellum* proportionality, my topic in the next section of this chapter.

be my focus of attention here. This is Brian Orend's formulation of the condition and it is as good as any:

> …[proportionality] mandates that a state considering a just war must weigh the expected universal (not just selfish) benefits of doing so against the expected universal costs. Only if the projected benefits, in terms of securing the just cause, are at least equal to, and preferably greater than, such costs as casualties may the war action proceed (Orend 2006: 59).

Going to war must be proportionate in that the total or universal costs, that is to say, the costs to *everyone* involved including the aggressor, must be weighed against the universal benefits, and the latter must at least equal the former.[9] Assuming that at least part, if not all, of the expected universal costs and benefits are the overall harms caused and prevented by going to war, then our justificatory principle JP applies.[10] This assumption is clearly warranted—wars harm and their only possible justification is that they prevent more harm—but when we try to apply the principle we find that weapons research introduces costs that cannot be measured and weighed. And it follows that any appeal to JWT to try to restrict the context of weapons research and so justify it, is in vain.

I said that *ad bellum* proportionality was not much discussed in the literature when I wrote *Designed to Kill* (Forge 2012) some years ago. Nothing has changed since. In particular, no one has challenged Thomas Hurka's explication, which was the starting point of my own previous discussion of the topic (Forge 2009). I argued that there are two distinct problems with *ad bellum* proportionality: what I called the *interpretation* problem and the *measurement* problem, a problem associated uniquely with proportionality. For example, as Hurka correctly points out, when faced with a condition like *ad bellum* proportionality we need to determine three things: (1) what are the benefits, or goods as he calls them, (2) what are the costs, and (3) how are these to be weighed against each other? (Hurka 2005: 38).[11] (1)–(3) may well look familiar, for they are generalisations of the three conditions which we identified as necessary for the application of JP: where Hurka refers to costs and benefits, we have talked about harms caused and harms prevented.

I raised questions about how one interprets costs and benefits in regard to answering (1) and (2), while dealing with (3) I took to be the substance of the measurement problem (Forge 2012: 211–213). This is not only a matter of trying to work out all the harms caused by war and all the harm prevented, and determining

---

[9]A theory about why states go to war which has it that states are only concerned with their own vital interests, such as Realism, would maintain that the calculation of costs vs benefits would be restricted to those of the state in question. They would be 'selfish' costs and benefits.

[10]If all there was to the costs and benefits were harms caused and harms prevented, then *ad bellum* proportionality would reduce to a specific instance of JP.

[11]These questions surely arise for anyone who would make some rational assessment about the decision to go to war, including those who are concerned exclusively with their own interests. Looking at the question from the Realist perspective, the problems are less difficult because it is only A's costs and benefits that are at issue, and not those of anyone else.

any other costs and benefits, because there is more than one way to proceed. For instance Hurka, and others, have argued that the benefits, the harms prevented, should only be those 'contained in' the just cause (e.g. Hurka 2005: 40). I think this is right—and it is the view that Orend expresses in the passage quoted above. What this means is that 'side-benefits' should not be included in any cost-benefit calculation. For instance, if in defeating A's aggression, B is able to gain a favourable trading deal with A's erstwhile partner D, then this benefit should not be included in the calculation.

If we refer to the selection of the benefits and costs as the dimension of the *scope* of interpretation, then we can also distinguish a dimension of *reach*, namely, how far into the future these are to be projected. Hurka view is "Although restricted in their content, the goods relevant to proportionality seem not to be restricted by their remoteness from a war or act either in time or causally" (Hurka 2005: 46). Moreover costs, on Hurka's account, are not restricted to those contained in the just cause "When we turn to the evils relevant to proportionality, we seem to find no restriction on their content parallel to the one on relevant goods" (Hurka 2005: 45). This is surely correct: there is clearly an asymmetry between costs and benefits here. One of the costs of war is re-building infrastructure—cities destroyed in World War Two for instance—and that cost is clearly not 'contained in' the just cause. Another cost of war is the effect of the health and well-being of the (surviving) population, such as was felt across Europe after World War Two. And another more remote cost was the fear engendered by the spectre of nuclear war as a consequence of the crash development of nuclear weapons and their use against Japan, in addition to the vast monetary cost of the nuclear weapons programmes since 1945. Benefits, such as modernisation measures needed simply because so much was destroyed will not, however, be counted in the calculation, as these were also not part of the just cause. The reason why the benefits are to be restricted to those 'contained in' the just cause is because it is those benefits and those benefits alone that justify the resort to war, they are what are *expected*. The only just cause for war is resistance against aggression and the benefits of such a war are the harms prevented that would otherwise be perpetrated by the aggressor. But the costs of war cannot be limited in this way. The costs of war are the price to be paid for resisting aggression and there is no non-arbitrary way to limit or circumscribe these. The costs of war do not abruptly come to an end when the war is over.

In the first section of this chapter when I considered some paradigm cases of harm prevention we agreed that a necessary condition, expressed by HC, for proceeding was to know how much harm was to be caused, how much pain a medical intervention would cause for instance. If that were not known, then there would be no reason to believe that more pain or harm would be prevented than would be caused, and the procedure would not be justified. Hurka's second condition, "What are the costs of war?", is exactly analogous. If one does not know what costs will follow from waging war, one is not justified in going to war according to JWT. But for wars of any significant magnitude and any significant duration, these costs cannot be known in advance, for the reasons just indicated. Once again, it is well-known that wars are the most risky and uncertain undertakings. If the costs and (some of the) benefits of wars do not become apparent until (long) after the war is over and if those costs are to be included in the total costs of the war, then the proportionality condition can only be seen to apply on the basis of a *retrospective* assessment of the war. What this means, in general terms, is that JWT is *not* a resource or guide for statespeople and others who decide whether or not to go to war. If one can only know if a necessary condition for engaging in a just war is satisfied until after the war is over, then the theory as a whole is useless as a guide for decision-making. But it does not follow that *ad bellum* proportionality should be abandoned as a necessary part of JWT. It is surely of over-riding importance that the costs of war should not outweigh the benefits, for if they did, then how could the war be just, or indeed the decision to fight rational? The fact that this cannot be known at the outset of the war means that no one can know at that time whether the war is just or not. There are implications here for weapons research.

At the end of the previous section I claimed that JWT and weapons research are not compatible, and in earlier work I argued that they are in fact incompatible (Forge 2012: 200). My claim was and still is that undertaking weapons research introduces costs, and possibility benefits, which cannot be determined in any estimate necessary for deciding if *ad bellum* proportionality is antecedently satisfied. In other words, weapons research is one of those things that makes the proportionality calculation impossible and implies that JWT cannot be a guide for statespeople and policy makers who are faced with decisions about whether or not to go to war. Weapons research is one of those particulars that makes the calculation impossible. Weapons research, like any other research activity, is uncertain as regards its outcome: will it be successful, how long will it take, when will the weapons be available? The outcomes of weapons research are uncertain: will they be available in time, will they be effective? The future of the outcomes of weapons research are uncertain: who will use the weapons, who will they be sold to, who will copy them? The future of the weapons research is uncertain: will this research lead to new weapons research and to new weapons systems? All of these uncertainties were exemplified by the Manhattan Project, and more examples will be given in the next chapter. Thus weapons research is not compatible with JWT because it vitiates the *ad bellum* proportionality calculus.

5.4 Conclusion

In order for an a harmful action to be justified it is necessary that it also prevents harm, for harming is morally wrong. And the harms must be comparable: there must at the very least be no more harm than there would have been had the agent not acted. A person who contemplates such an action must therefore have a reliable estimate for how much harm her act will cause and how much it will be prevent, and be able to compare the harms caused and harms prevented, *before* she acts. We have seen that even in the most controlled and best understood cases in institutions set up to prevent harm, in medicine for example, it is often hard to make the right calculation. One of the problems in general here is estimating the harms prevented, for, by definition, these do not or will not exist when the action is undertaken. But first it is necessary to work out the harms caused, as stated in HC. If this first step cannot be completed, then there is no hope of justifying a harmful action. I considered whether it might be possible to side-step this problem by restricting the context of weapons research to democratic states or to states fighting a just war, and so identifying at least some possible instances where weapons research might be justified, but to no avail.

Appendix: Pacifism and Weapons Research

JWT has been compared and contrasted, by Orend for instance, with Realism on the one hand and Pacifism on the other (Orend 2006: 4–5). The difference between the three in regard to war is (roughly) as follows: JWT holds that war is morally permissible under certain conditions, namely, those set out in the theory; Pacifism denies that war is ever morally permissible, while Realism maintains that morality has nothing to do with wars nor any other kind of relations between states (cf Carr 1946: 153). Pacifism and JWT resemble one another in applying a moral standard to war, but differ in that the former's is stronger. For the Realist, war may be necessary or expedient, depending on whether it defends or promotes the vital interests of the state, nothing more. My own view is that Realism is the correct account of what wars are and why states go to war, whatever statespeople might say. I have given no argument in favour or against either of the other two theories as the correct normative account of war, and, moreover, as should be clear by now, my argument against weapons research does not presuppose either theory, in that it has no premise drawn from either of them. I have focused on JWT because it is more permissive in regard to war and hence one would suppose it would allow (some, or more) weapons research. We have seen that it does not. My argument against weapons research is therefore consistent with JWT, but does it give us any reason to prefer it to Pacifism, or conversely?

Pacifism and JWT evidently differ in terms of what roles they allow for armed forces and the weapons they wield: JWT allows that armies, etc., can go to war to resist aggression, but Pacifism in the traditional sense does not. This in itself, however, does not imply that Pacficists *must* maintain that states cannot have armed forces, it is just that they cannot use them to fight wars. This might seen spurious until we remember that weapons can be used to deter, to prevent others starting wars. A pacifist deterrent threat will be empty because war is not permissible, but that does not mean that it could not be efficacious. In this way we can see how Pacifism can be consistent with weapons research, namely with weapons research done with the intention of providing the means for deterrence. We have established that deterrence is a derivative function of a weapon and that there is no such thing as 'purely deterring' weapons, and hence any pacifist-sponsored weapons research would still have to produce the means to harm in order to obtain the means to deter.

One might expect Pacifists to also maintain that threats to violence are impermissible and hence that deterrence is also not allowed. Nevertheless, it appears that Pacifism *simpliciter* and weapons research are strictly speaking incompatible, in the sense that the denial that war is ever morally permissible does not imply that weapons research is morally impermissible.

There are versions of Pacifism that do not deny that participation in all wars is morally wrong, something that Orend for one finds 'strange' (Orend 2006: 244), with one such variety, known as contingent Pacifism, being of current interest (for example, May 2011; McMahan 2010). It is worth saying a little more about this idea here. The point about contingent Pacifism is that its starting point or main premise is not that war is always forbidden. It is rather than wars are very bad indeed, and should only be undertaken if it is clear that they will prevent outcomes that would be very much worse than the war is. This is a stronger demand that there is just cause and that the proportionality condition is satisfied.[12] However, and here we do see some resemblance to my position, since one can never be sure that this demand is satisfied, in practice it entails a prohibition on all war, hence the Pacifist stance is *contingent*. If there were a complete cessation of weapons research, given that everyone were to accept that it is morally wrong, then this would eventually lead to world in which it would look as if everyone was a Pacifist because there would no longer be the means to fight wars, so no one could fight and there would be no more war. This not contingent Pacifism as such because the argument does not depend on any judgement about the rightness or wrongness of war. However, since the contingent Pacifist holds, in effect, that the *ad bellum* proportionality calculation cannot be carried through, then she and I share the same view on this issue.

References

Brody, B. 1988. *Life and Death Decision Making*. Oxford: Oxford University Press.

Carr, E. 1946. *Twenty Years of Crisis*. London: MacMillan.

Forge, J. 2009. Proportionality, Just War Theory and Weapons Innovation. *Science and Engineering Ethics* 15: 25–38.

Forge, J. 2012. *Designed to Kill: The Case Against Weapons Research*. Dordrecht: Springer.

Gert, B. 2005. *Morality: Its Nature and Justification*. Revised Edition. Oxford: Oxford University Press.

Hurka, T. 2005. Proportionality and the Morality of War. *Philosophy and Public Affairs* 33: 34–66.

May, L. 2011. Contingent Pacifism and the Moral Risks of Participating in War. *Public Affairs Quarterly* 2 (25): 95–111.

McMahan, J. 2010. Pacifism and Moral Theory. *Diametros* 23: 48–68.

Orend, B. 2006. *The Morality of War*. Peterborough, Ontario: Broadview.

Walzer, M. 1977. *Just and Unjust Wars*. London: Allen Lane.

---

[12]This is reminiscent of Walzer's notion of a supreme emergency, see Forge (2012, Chap. 13).

Chapter 6
The Changing Contexts of Weapons Research

6.1 Introductory Remarks

Weapons research is conducted at particular times and places, under certain circumstances and conditions. I have expressed this by saying that weapons research
takes place in some *context*. There have been instances of individuals conducting
weapons research simply out of interest—Hiram Maxim the inventor of the first
effective machine gun, fits the mould of someone who just liked inventing things—
but in the vast majority of cases it is conducted in response to something external,
something outside the research facility.[1] Since weapons research aims to produce
weapons and weapons are the means to fight wars, and wars are the business of the
state, it is the norm for states to commission weapons research. And they do so, on
the whole, on the basis of reflection of what they think they need by way of
armaments, and that in turn is a function of context. When Germany attacked the
Soviet Union in 1941, the context in which the latter found itself suddenly changed
and it realised that there was an urgent need for as many tanks as it could get. Recall
that Martin Wight has told us that what a state's security needs are are. what it
perceives them to be, and while something of a truism, this does imply that these
are subjective to a degree. So in general we can say that it is the norm that weapons
research is commissioned by the state in response to its perceived security concerns.
But the security needs of a state can change and this means that it may require new
weapons and that the weapons previously commissioned are no longer adequate. In
fact, the state itself can change, as happened to Nazi Germany in 1945 and the
Soviet Union in 1991, leading to a dramatic reappraisal of matters to do with
security. In the first section of this chapter I will make some more comments about
the nature of contexts and how they influence weapons research.

---

[1]Maxim is on record as saying that when firing a gun at age 26 and feeling the recoil, he wondered
if this 'work' could be used to fire another round and so produce an 'automatic' weapon. Many
subsequent generations of automatic weapons have used this principle, see Forge (2012: 67–71).

© The Author(s), under exclusive licence to Springer Nature Switzerland AG 2019   77
J. Forge, *The Morality of Weapons Research*, SpringerBriefs in Ethics,
https://doi.org/10.1007/978-3-030-16860-5_6

We have seen that the outcomes of weapons research in the form of designs outlive the contexts in which they were created. And they can be reproduced without limit wherever the designs are available, provided that there are the means and motivation to do so. Thus as the context changes in ways which impact the security needs of states, so the products of weapons research can come to be used in ways that could not have been anticipated at the time the work was done. If follows that the weapons researcher cannot know how much harm her creations will cause because she does not know how they are going to be used, and that is because she cannot know, among other things, the form of the security issues in the new context. In fact it is often the case that no one can anticipate these things—the abrupt ending of the Cold War can again be mentioned. Weapons, and weapons designs, can be transferred between countries, by license, sales and even theft, where they can come to be used in ways different from the intended uses in the originating country, and this also amounts to a change of context. That the outcomes of weapons research live on through changing historical periods is thus an important reason why weapons researchers cannot know what harms their creations will cause. The three examples I will give in what follows all illustrate and confirm this claim. And we should not forget that even within what is considered to be one and the same context, the ways the output of weapons research are used and the harms they cause cannot be accurately estimated. The Manhattan Project lasted almost as long as the US participation in World War Two, but none of those taking part could have anticipated how the bombs they created would be used, and neither could anyone else have done so right up until the successful test in July 1945.

One might expect that the most pressing need for weapons will come about when there is a war—the Manhattan Project took place in World War Two. One of the examples to be considered here, the Kalashnikov or AK-47, also has its origin in World War Two. The AK-47 came into production in the context of an emerging superpower, not a country fighting for its very existence. Moreover, it found its way to many countries and was used in many wars, and so the example illustrates two of the ways in which contexts can change. Another source of weapons research, especially in more recent times, is military doctrine. I will give an example of how a change in military doctrine in the US led to the development of a range of new weapons. In this example also the context changed before the weapons could be used for the purposes for which they were intended. My first example concerns the evolution of the rifle in the nineteenth century. The example differs from the other two in that I want to use it to illustrate the way in which weapons evolve; how one weapons innovation leads to another. This is a feature of many kinds of design and artefact invention, that one innovation suggests another and that 'generations' of artefacts form 'lineages'. This demonstrates another way in which weapons research projects into the future: it has descendents. One need not accept this 'evolution view of technology' as a whole to agree that it is often the case that one innovation leads to another and that without the former, the latter would not have come about.[2] A final comment: the reader may already be convinced that weapons research cannot be justified, in which case it is gratifying that I have achieved my objective, but I think this chapter is still important.[3] It is necessary to make as strong a case against weapons research as possible, bearing in mind all the attempts that are made to vindicate defence spending and rationalise military ventures that we are accustomed to.

6.2   Contexts

So far I have talked about the context of weapons research informally, as the specific or particular time and place where the research takes place, and have said that certain prevailing circumstances and conditions are instrumental in bringing about the research in question. This intuitive idea may be enough to work with, and I think it is clear that weapons research does not take place in a vacuum. Nevertheless, it is appropriate here, before we turn to some more examples, to try to be a little more precise about what contexts are by amplifying some of the remarks made above. There are two matters we would like to know about: what or who decides the weapons research projects that take place in context **C** and what determines how the products, the weapons, are used in **C**. The second question is important as well as the first. If security concerns of the state change within **C** and it is decided that weapons commissioned for one purpose are now needed for another, then that will not be something (easily) anticipated at the time the weapons research was conducted. The third example illustrates this possibility. If **C** itself changes and the state finds itself in radically new situation with new security concerns, then there will be still greater lack of transparency, as we will see from the other examples. And finally if the state itself disintegrates, as the Soviet Union did, then this will transform not only **C** but also perhaps the contexts of other states and creates still greater difficulties in predicting how weapons will be employed and in making estimates about harms caused by them.

Posen's view of the grand strategy of a state and the corresponding military doctrine was quoted Chap. 2, and I have referred to these ideas a number of times. Since military doctrine is concerned with military means, it is concerned with what weapons are required to carry out any military action needed and hence, as we have remarked, it is concerned with weapons research. So military doctrine is clearly a *contextual factor* that is relevant to weapons research. Is it the case then if military doctrine changes, then so does **C**, or rather, is this the best way to understand what the context of weapons research is, to say that it is defined by military doctrine? An alternative, also suggested by Posen's remark, is that **C** changes when the state's

---

[2]This is a characteristic of all forms of research—later work is informed by earlier work, one achievement leads to another. For more on this, see Basalla (1988).

[3]The reader may be excused from wading through this chapter, in that case.

grand strategy changes, so on this reading it is grand strategy that defines the context. My preference—there is no right answer—is for the latter proposal simply because the idea of an (explicit) military doctrine is a relatively recent one, whereas ever since there have been states, these have worried about their security. Thus the beginnings of military doctrine in Posen's sense can be dated from the nineteenth century, shortly after the end of the Napoleonic Wars, when Prussia set up institutions dedicated to planning for war and training officers, the origin of a general staff. Henry the Eighth, who embodied the English state, did not have any general staff or explicit military doctrine, but he knew the French were his enemy and he knew he needed ships and gunpowder weapons to fight them, so he did his best to equip his army and navy with the best available. Other kings and rulers who, like Henry, guided the destiny of states with the aid of small groups of trusted advisers prior to the nineteenth century, had grand strategies in the sense that they knew who their enemies were and tried to work out the best ways to deal with them. A state's grand strategy is not necessarily exclusively concerned with its security, and security is not necessarily always a matter of acquiring weapons, but the acquisition of weapons is, on the whole, a response to grand stategical considerations.

Within the context of one and the same grand strategy, military doctrine can change. For instance, the Soviet Union was perceived as the main security threat for the United states from about 1947 (and vice versa). This remained true until the end of the Cold War. However, the ways in which the United states tried to address this threat, in terms of nuclear military doctrine changed, from Massive Retaliation in the 1950s to nuclear deterrence or Assured Destruction when the Soviet Union achieved nuclear parity in the 1960s. The latter doctrine called for survivable nuclear forces and so led to the development of nuclear submarines, which were not needed to massively retaliate, but the bombers and ballistic missiles previously intended for this purpose became part of the 'triad' of nuclear deterrent forces. On the other hand, military doctrine, or elements thereof, can remain in place while grand strategy changes, as we will see. Finally, grand strategy and military doctrine can change together. Hitler's grand strategy was to expand German possessions and influence over much of Europe. Before 1933 when Hitler assumed power, Germany was highly vulnerable and surrounded by enemies, which it was not in a position to do anything about—it was in a grand strategical vacuum with a very small army and much diminished armaments industry. To carry out his grand strategy, Hitler needed an offensive military doctrine, a large army and an armaments industry, all of which he got.

There is another important contextual factor which lies at the other end of the spectrum from grand strategy, and that comprises the military technology already available in **C**, the weapons that already exist. Much research activity involves making use of and building on previous achievements, and weapons research is no exception. I mentioned the evolution view of technology, which sees artefacts as forming lineages, with each successive generation being an improvement on the previous one and said that one of the examples to be considered illustrates this theory. Again, we do not need to accept the theory to accept that previous work is often a necessary step in the development of new systems. There have not been too many really radically new kinds of weapons, although we have seen some examples thereof—the chariot, torsion artillery, gunpowder weapons and nuclear weapons were radically new, and gave rise to new 'lineages' of weapons.[4] So while we can now think of the process of weapons innovation and development as a response to the demands of grand strategy for a perceived need for the means to address the security of the state, this response can only be *expected* to be based on what is available, not to conjure up something radically new.

6.3 From Musket to Rifle

At the beginning of the nineteenth century the standard infantryman's weapon was the musket. All the armies that fought in the Napoleonic Wars were equipped with muskets, as were soldiers in North America, in China and Japan, in the armies of the Ottoman Empire, colonial troops in Australia and Africa, and more besides. There were some troops equipped with rifles. These were sharpshooters or skirmishers who were separate from the ranks of infantry who would advance in massed lines or columns to discharge their muskets, preferably all at the same time, at a distance of less than 100 m. Rifles have rifling—hence the name—spiral grooves and 'lands' which run down the inside of the barrel and ensure that the bullet spins along the axis of the barrel. This makes them more accurate than muskets, which are smoothbore.[5] But all the available gunpowder weapons, muskets, cannon and rifles were muzzle-loaders and all the available ammunition were iron balls of various sizes, which meant that the ball had to be hammered down the rifle barrel against the lands, which was slow and destructive to the weapon. Also, the gunpowder was black powder, which produced a cloud of sulphur dioxide when fired and left a residue of carbon and sulphur products in the barrel. The latter fouled both muskets and rifles but was much worse for the latter than the former. Although rifles were more accurate than muskets, they were slow to reload and quick to foul, which was why their use was restricted to specialised soldiers.

Firearms were invented in China and were considerably improved in Europe. Gunpowder was also invented in China, in the 800s, and there is a formula for the explosive recorded in a military text, no doubt the product of weapons research, dated from around 1040 (Chase 2003: 11). Depictions of the first gun are from the

---

[4]One might see these, to use the language of Thomas Kuhn, as akin to new paradigms in science.

[5]The combination of a smoothbore barrel and round shot, be it fired from a musket or a cannon, renders the weapons inherently inaccurate owing to the uncontrollable spin imparted to the projectile as it leaves the barrel. Unless by good fortunate the ball is spinning along the axis of the barrel, it will be deflected in the direction of the spin.: for an explanation with reference to the physics, see Forge (2012, Footnote 5, Chap. 4). Kenneth Chase, whose work I shall refer to in the next paragraph, points out that a skilled archer could fire arrows much quicker than a musketeer, had a much better chance of hitting a target, and could do so at longer range. To read why the musket nevertheless prevailed, see Chase (2003: 73–74).

twelfth century. Firearms appeared in Europe in the 1300s, almost certainly brought by the Mongols who had conquered all of China by 1276 and so had access to guns from the Chinese (Chase 2003: 58). The original Chinese formula for gunpowder was not optimal in terms of the proportion of the three ingredients, carbon, sulphur and potassium nitrate, and neither was the method of combining them, which was simply to mix them up. This meant that fairly a large amount of powder was needed and this entailed heavy guns: the first important gunpowder weapons were massive cannon. Changes to the proportion of the ingredients and a method of combination gave rise to more powerful explosives: grinding the ingredients gave a finer mixture known as serpentine powder, while corning, where liquid was added to the mixture and allowed to dry forming small grains known as corns, provided a more powerful form still—both the result of weapons research.

These innovations took place in the fifteenth century and they enabled effective smaller guns, the first of which was the arquebus (Chase 2003: 61). Guns known as muskets, or Spanish muskets, were also invented, which were heavier versions of the arquebus and required a tripod when fired. All the armies of Europe were equipped with gunpowder weapons by the end of the fifteenth century, as were the Ottomans in the East, and at the beginning of the next century, European weapons were introduced back into China by the Portuguese. The weapons spread by a variety of means, by trade, by capture, by imitation and by the skilled gunsmiths who travelled round Europe selling their trade.[6] The original arquebus was fired by setting a match to a touch hole to ignite the powder. This was an awkward, time-consuming process which often led to a misfire. The next innovation was the flintlock mechanism, whereby a spark from a struck flint ignited the powder, and which gradually replaced the matchlock. Flintlock muskets became available in the seventeenth century and all the armies in Europe possessed these by the middle of the eighteenth. The British army, for instance, used the same 'Brown Bess' musket from 1722 until the middle of the nineteenth century. The weapons and tactics of European armies did not change much either during the period just mentioned. Cavalry, artillery and infantry were the types of formation employed on the battlefield, and when infantry engaged with infantry, the protagonists would march up to each other, either in line like the British or in column like the French, and start to fire at 100 yards (91 m). The soldiers had to stand up because they could not reload otherwise. After several volleys the battlefield would be shrouded in smoke and then one side would charge with bayonets.

The chances of being hit by a musket ball fired at a range of 90 or so metres was pretty slim, as I have said, otherwise the infantry would not have survived. In the nineteenth century several innovations, starting in the 1840s, rendered these tactics suicidal. There were a number of attempts dating from 1818 to invent a bullet that

---

[6]For instance, Peter van Collen, a leading Belgium gunsmith, was brought to England by Henry VIII to set up gun shops to increase the supply and stockpile of weapons available to the English armies. These were established around the Tower of London where the royal arsenal was located, and Henry encouraged English gunsmiths to set up there as well.
DOI http://firearmshistory.blogspot.com/2010/05/barrel-making-early-barrel-making-in.html.

did not have to rammed down the barrel of a rifle. Claude-Étienne Minié, a captain in the French chasseurs who served in Africa, did so successfully in 1849. He designed a conical lead bullet which was smaller than the bore of the rifle or rifled musket, so was quick to load, but was so designed that an iron cup in the base expanded when the rifle fired to enlarge the bullet so it was caught by the rifling and the spin imparted (McNeill 1982: 231). The increase in accuracy was considerable, with firearms using the bullet estimated to the eight times more accurate than the musket. The Minié bullet was indirectly responsible for changing the tactics of infantry warfare, for no longer could infantry advance in line or column and deliver volley fire, and it also heralded the end of cavalry as an independent formation.[7] Two observations about Captain Minié before we move on: first of all, the chasseurs were the skirmishers of the French army, elite troops who traditionally used rifles, so although we do not have a full account of how and why he came to invent his bullet, we can surmise that it was a response to the problems inherent in the design of the muzzle-loading rifle. The second is this: the Minié bullet, was designed by a Frenchman, a member of the light infantry who served in Africa in the years of French colonial expansion, and was then subsequently used to terrible effect for twenty years in wars in continental Europe and America that Minié never served in and could not have anticipated.

The Minié bullet posed a problem for infantry tactics: it was accurate, so troops needed stay down under cover, but on the other hand it was necessary to stand up to reload. In the first war where the new bullet was used, the Crimean War, only one side had it so the problem did not arise. It did in the American Civil War where both sides were armed with Minié rifles and muskets (Lee 2016: 366–368). We could sum up Minié's achievement by saying that he actually rendered muzzle-loading weapons unsuitable for war by making them too accurate. There would no longer be a problem if it were possible to re-load while lying down, and this was made possible by breech-loading rifles, which gradually replaced the musket and Minié's revolutionary bullet, during the nineteenth century. In fact, an effective breach-loading rifle was invented fifteen years before the Minié bullet and was adopted by the Prussian army in 1840, but relatively few were purchased. This was the Dreyse needle gun, invented by Johann von Dreyse (McNeill 1982: 235). It was called a needle gun because the firing principle involved a needle penetrating and igniting a paper cartridge, which then fired the bullet. The cartridge and bullet were inserted into the breech by pulling back and then engaging a bolt. The gun was simple to use, the needle could be easily replaced and it had a good rate of fire. But the musket still had one advantage: it did not leak gas from the breech the way the Dreyse did. The other disadvantage of the Dreyse, one it had in common with the musket, was that it was a single shot: the soldier had to reload, as he always had in the past, after every shot. Dreyse himself

---

[7]This lesson was not learned by Pickett's division at Gettysburg in 1863. Robert E. Lee ordered the division to charge the Union forces on Cemetery Hill. At a range of about half a kilometer the Union troops opened fire and killed over half the men, using Minié bullets.

learned to make guns and ammunition as an apprentice in Paris to Jean-Samuel Pauly, an eccentric Swiss who, *inter alia*, designed several (ineffective) breech-loading guns.

From 1860 to the end of the century, all the major European powers, Russia, Japan and others developed highly effective breech-loading rifles firing cartridges filled with smokeless power. The rifles had similar performance and differed only in capacity of the magazine and the calibre of the bullet. These innovations owed much to the application of science to their production, both in regard to the steels using in their construction and their ammunition. I will mention one of these, the German Mauser, in a little more detail (and see Grant 2015). The development of the Mauser began in 1866 with Paul Mauser's work on the Dreyse, when he developed a self-cocking system. This meant that the needle mechanism did not need to be retracted before the bolt could be opened, as with Dreyse's original gun, pulling the bolt back, adding the cartridge and pushing the bolt forward readied the gun for fire, saving one action and increasing the rate of fire. Mauser next adapted the rifle for a self-contained metal cartridge. In 1871 this rifle, called the *Infanterie-Gewehr Modell* 1871 was formally adopted by the Prussian army. It was something of a *mélange*, with the barrel and rifling design stolen from the French Chassepot, the trigger unit from the 1862 Dreyse, with Mauser only responsible for the bolt. After some years of work, Paul Mauser developed a rifle with a five cartridge magazine which pushed a new cartridge up into the breech as the bolt ejected the spent cartridge, with a stripper clip to hold five cartridges for reload. This was the *Gewehr* 88, a model that also used smokeless powder cartridges. The final model, the *Gewehr* 98 comes with this distinctive clip, but was greatly improved in other respects over the 88.[8] Similar stories can be told of the development of the rifles of the other major powers, with individuals working on ways to improve the inventions of Minié and Dreyse. These rifles became the standard infantry weapons for the next half century and beyond.

The Mauser 98 and its shorter barrelled carbine the 98b were used in both world wars. In the Second World War it was standard weapon of the *Einsatzgruppen* who followed the Wehrmacht into the Soviet Union to exterminate the Jewish population (MacLean 1999: 14). Neither Minié nor Dreyse nor Mauser, nor any of the other weapons designers who worked on rifles, could have anticipated or even imagined the nature and scale of the conflicts of the twentieth century. But without their work Germany, Britain, France, etc. would still have been fighting with muskets, and many lives would have been saved. The point of this discussion has been to indicate how weapons research feeds off previous work, how one innovation leads to another, how weapons researchers seek to deal with the problems and shortcomings of the present generation of weapons. Much more could be said about the evolution of the rifle from the musket. The moral of the story is that the effects of weapons research are not only direct, in that it aims to produce a particular weapon which causes harm, but also indirect in that it contributes to future generations of weapons as well. And this is what usually happens, as we have seen in

[8]For more still, see Forge (2012: 62–67).

other examples mentioned in this book. For instance, the war cart led to the chariot, the bow to ancient artillery, atomic weapons to thermonuclear weapons and so on. Only a very few weapons were created without some basis or model to provide inspiration—the invention of gunpowder and the atomic bomb were two exceptions.

I mentioned that according to the evolution theory of technology, the development and progress of technology can be thought of as series of generations of artefacts, with later generations being improvements, in some sense, on earlier ones, and with the latter leading to the former. It therefore seems appropriate to speak of generations of muskets, rifles and ammunition, with a continuous improvement in performance, and in the majority of cases with one innovation being the basis for the next. What I mean by the indirect effects of weapons research are the harms that weapons produced by episodes of weapons research that use the previous generation of weapons as a basis or model or inspiration. Paul Mauser used the Dreyse rifle as his first model, seeking to deal with gas leaking from the breech, then his models were his own designs which he sought to improve. The direct effects of Dreyse's work was the harm caused by his rifle, for instance during the Austro-Prussian War of 1866; the indirect effects included the harms caused by the *Infanterie-Gewehr Modell* 1871 in the Serbo-Bulgarian War of 1885. There were thus three things that Dreyse, to take one instance, could not know when he designed the needle gun: he could not know where and when it would be used and how many people it would kill; he could not know how, where or when it would be used as a basis for an improved rifle, and he could not know the extend of the harms caused by this next generation of rifle. Moreover, there were more generations of rifles to come, in this 'lineage', and much more harm caused.

We have seen that Mauser's research began with his work on the Dreyse rifle. It is possible that Mauser would have designed the *Infanterie-Gewehr Modell* 1871 without the help provided by Dreyse, but as a matter of fact he didn't and we cannot know if he could have done so. Having Dreyse rifles to examine was one of the *causes* which led Mauser to his first successful innovation. What I understand here by cause is that having Dreyses was a *necessary condition* for Mauser's successful work in the sense that, as a matter of fact, focussing attention on the Dreyse needle gun was one step on the path than led to the *Infanterie-Gewehr Modell* 1871. There is thus a clear sense in which Dreyse's work had indirect effects: it was, as a matter of fact, something that contributed to the invention of the Mauser rifle, which in turn killed people in the Serbo-Bulgarian War. Without the Dreyse to work on, Mauser could still have invented his rifles, but, as I have remarked, we cannot know that. The unknowable future harms brought about by weapons research are thus compounded by indirect effects, and hence the possibility of justifying weapons research is pushed still further out of reach.

6.4   Mikhail Kalashnikov: AK-47[9]

The legendary *Avtomat Kalashnikova* 1947 assault rifle, or Kalashnikov AK-47 as it is commonly known, was invented in 1946 by Mikhail Kalashnikov—according to folklore at any rate.[10] It was issued to the armies of the old Warsaw Pact countries as well as China and used in many conflicts, by the North Vietnamese Army (NVA), by both sides in the Soviet invasion of Afghanistan, by Al Qaeda operatives in Iraq and Taliban forces in Afghanistan and Pakistan, and by liberation movements and later child soldiers in Africa and South America. It has been by far the most widely produced weapon in history and by far the most widely used since the Second World War, with at least sixty million units being produced. Some design modifications have been made since the first models were made and new versions of the AK have appeared periodically (ironically, the AKs now fire the standard NATO round).[11] In his old age, Mikhail Kalashnikov came to have some doubts about his invention. He told *The Times* in June 2006 "I don't worry when my guns are used for national liberation or defence. But when I see how peaceful people are killed and wounded by these weapons, I get very distressed and upset. I calm down by telling myself that I invented this gun 60 years ago to protect the interests of my country." It was not the case, of course, that Kalashnikov was concerned about numerically the same weapons being used for national liberation (NVA for instance) and for killing peaceful people (Shiites for example)—one assumes the weapons used by the NVA are by now rusted and useless—it is rather than they are weapons made according the design or blueprint that he invented back in 1946.

In his book, *The Gun that Changed the World*, Kalashnikov tells the story of the invention of the AK-47 in his own words (and that of his co-author).[12] The story is very much rooted in the Great Patriotic War, the Soviet name for the Second World War, and told no doubt with the benefit of hindsight. Kalashnikov was probably a talented inventor, remarkably so as he had no technical training and not much

---

[9]Discussion here follows Forge (2012: 71–75).

[10]This statement should be qualified, in view of some confusion and myth surrounding this iconic weapon. It is certainly more accurate to say, at the later stages of development, Kalashnikov was only part of the team that perfected the design. Such historical details are, of course, important, but what became of the gun after its invention and how this supports the proposition about the lack of control of weapons designers, whoever they are, and the unpredictability of future uses is even more so. In what follows I give the 'standard version' of the story. Of course, the story of a 'simple' soldier from the Great Patriotic War designing a gun to help defend the Motherland accorded well with Marxist-Leninist ideology—material need bringing forth the technical means to satisfy that need by the proletariat—and is also a good story.

[11]We have here another illustration of the idea that there are generations of weapons, that weapons come in lineages. Not only have there been different models of the AK, but there have been submachine guns based on the design. See Shilin and Cutshaw (2000) for details.

[12]Here I have used Kahaner (2007), Ford (2005), Bishop (2006) and Chivers (2010). Much of the material is standard and widely available, so I have not included references at every stage. Chivers (2010) is the most recent source I have seen, and the best.

schooling. Kalashnikov's war ended quickly as far as fighting was concerned: he was wounded in his tank at the battle of Bryansk in the fourth month of the war and that was the end of his combat role. As he was trying to make his way back to his own lines with wounded comrades, he heard bursts of German gunfire and was told it was from machine pistols, so the story goes. Shortly afterwards he came across wounded Soviet soldiers who had been subsequently killed with the machine pistol, the *machinenpistole* 40 or MP40. In hospital he read all he could about machine guns, and then, on being discharged, got himself sent to Kazakhstan, where he got access to a machine shop in a railway yard, and he was actually able to produce a sub-machine gun out of spare parts. This was in 1942. Kalashnikov then got himself sent to Uzbekistan to work with an automatic firearms specialist Anatoli Blagonravov, then to the Shurovo Polgon weapons inventions department in Moscow. Here Kalashnikov designed another machine gun, but his design was not selected for mass production. As we know, AK-47s appeared on the scene too late for the war. In fact, Kalashnikov did not turn his hand to designing assault rifles until 1945.

The Germans were the first to make an assault rifle, the *machinenpistole* 43 and the *sturmgewehr* 44.[13] The difference between an assault rifle and a rifle on the one hand and a (sub) machine gun on the other is that the former can fire both bursts of bullets and single shots. Like the rifle it carried a magazine, but unlike the rifle, the magazine normally holds 30 rounds. Technical advances in metal pressing and stamping meant that the Germans could mass produce the MP43 or *machinenpistole* 43 also called the Sg44 or *sturmgewher* (assault rifle) 44 and intended to equip all their infantry with them—the first models went to the elite SS troops and to then to the soldiers fighting in the East—replacing the Mauser as standard equipment. Another German innovation was the use of smaller ammunition. The MP40 fired 9 mm bullets, but the MP43 used a 7.92 mm bullet, continuing the trend to smaller rounds noted earlier. This was less powerful, but lighter and hence caused less wear and tear on the gun and was easier to fire, a big advantage when automatic mode is selected. Recall that it is not so much the mass of the bullet that is important when it comes to damage inflicted, but its velocity.[14] In 1974, before the invasion of Afghanistan, the model AK-74 was released, this time with a 5.54 mm bullet. This had an airspace or hollow cavity—effectively, a dum dum round—which made the bullet deform on impact and so give up most of its energy before exiting the body. The Mujadiheen, who suffered from this new model, noted the magnitude of the

---

[13]The story is that leading German small arms designers, Schmiesser, who invented the MP40, and Walter wanted to make an assault rifle and the High Command also wanted one, but Hitler liked machine guns and sub-machine guns. Hence they disguised their first assault rifle by calling it the MP43. It was such a success that Hitler himself coined the name "sturmgewehr". Schmiesser was captured by the Soviets and send to Izhevsk, the city in the Urals that was the design, development and manufacturing centre for Soviet small arms, and he was there when the final stages of the AK-47 development. It would be strange indeed if he had not been involved.

[14]The muzzle velocity of the Sg44 bullet was 650 metres per second versus 776 metres per second for the Mauser, quite enough to damage its victim.

injuries, and dubbed this the 'poison bullet'. The assault rifle has now become the standard infantry weapon, replacing the old-style rifle, but even after sixty years, the AK is still one of the best of its kind.

The AK was taken up by the USSR as the main infantry weapon in 1949, the same year as the foundation of NATO. Soviet doctrine at the time taught that a war with NATO would most likely involve massive conventional engagements with tanks and massed infantry battles. The relatively poorly trained Soviet conscripts would be able to sustain very high rates of fire with the AK against their probably better trained (but in fact not better equipped) enemies. The gun was also to be used by the USSR's Warsaw Treaty Organisation allies, after the WTO was formed. It would obviously be simpler, cheaper and much more practical if all of these forces were equipped with the same weapon and the same ammunition, so after Stalin died, Khrushchev was willing to export it to his allies. Not only that, it was licensed for manufacture, free, to Poland in 1956, Hungary in 1958, East Germany and Bulgaria in 1959 and later to Romania. But it was not just WTO countries that were allowed to make the AK: it was also licensed to China in 1956, to North Korea in 1958 and to Yugoslavia in 1964. Other countries that have, or still do, manufacture the AK include Iraq, Egypt and India. Variants of the AK are made in Finland, Israel and South Africa. It is clear, then, that the spread of the technology is completely out of control. Moreover, the spread of the gun is also out of control, as it is easy to buy, and cheap. Movements and groups that have bought the gun in include: Viet Cong and NVA, Sandinistas, FARC, Mujahideen, Taliban, Al-Qaeda, Frelimo and the NPLF.[15] Wars have been fought with AKs as the main, or the main infantry, weapon on almost every continent. There has also been the spread of what has been called 'Kalashnikov Culture', a term used to describe the central role of the gun amongst certain subgroups in the developing world for control and extortion, a far cry from Mikhail Kalashnikov's dream of producing a gun to compete with the MP40.

Mikhail Kalashnikov tells us that he invented his gun to 'protect the interests of his country'. But the context in which the gun was invented, or rather, the context in which (the story has it) the idea of an assault rifle (something) like the AK-47 was conceived, is totally different from the contexts in which it has been used. The Soviet Union was facing literally a struggle for its very survival as a state in 1941, when Kalashnikov was wounded.[16] In 1949, when the gun became available, matters were very different. The USSR was a superpower, which had invented a nuclear weapon, had the largest army in world and had installed communist regimes

---

[15]The spread of the AK is an instance of the diffusion of technology in which the way the technology is used is by group B, who take up the technology, is considerably different from the way it is employed in the 'home country'. It was never intended, or expected, to be used by child soldiers in Africa and South America. This again illustrates the unpredictability of weapons research: the way in which weapons technology diffuses is uncertain.

[16]Or, more strongly given the Nazis' intentions, the Russian peoples were in a struggle for life or death.

in seven European countries. The first actual uses of the gun were for repressing dissent, notably the Hungarian Uprising in 1956. Here is Chivers' scathing assessment

> The AK-47 was christened with blood [in 1956] not as a tool of liberation or to defend the Soviet Union from invaders. It made its debut smashing freedom movements. It was repression's chosen gun, the rifle of the occupier and the police state. The beginning established a pattern. The Kalashnikov was rarely [ever?] a Soviet weapon of defence. It was the weapon of East German border guards who shot unarmed civilians fleeing to the West...it would be used in Prague, Alma-Ata, in Riga, in Baku...in Tiananmen Square, in Andijon in Uzbekistan and Bishkek in Kyrgyzstan – almost any place where a government resorted to shooting its citizens to try to keep [them] in check. (Chivers 2010: 220).

The Hungarians who resisted the Soviet troops sent to end their uprising captured some AK-47s and used them in turn. This was the first occasion, but not the last, that the gun was used against Soviet or Russian forces, or by liberation movements against other repressive regimes or invaders. If repressing freedom movements or invading countries, Afghanistan for instance, is bad, then it seems that resistance is good, as are the means used for resistance. In other words, set against the 'bad uses' of the AK-47, there are some 'good uses'. But the point of this example is to show, first of all, that the context in which is was conceived and invented and that the purpose for which it was intended, changed and hence that, secondly, the task of determining the harms caused by the AK is simply impossible. AK-47, was never used to fight against the Nazis, in fact, Nazi Germany had ceased to exist before the first test of the gun was ever made.

6.5   The Changing Context of Military Doctrine: The 'Big Five'

In the last section I said that the Soviet Union equipped its troops and those of its Warsaw Pact allies with AKs because it thought that this most likely confrontation with NATO was a conventional war in Europe. And the US, and its NATO allies, came to agree. Recall that in Chap. 4 we discussed US grand strategy and nuclear doctrine in the 1950s, and saw how containment and the corresponding doctrine of massive retaliation became untenable when the Soviet Union developed survivable nuclear forces and then achieved parity—Soviet nuclear doctrine, set in train by Stalin in 1946, had made this the top priority. Both superpowers believed that the other was aggressive and would destroy its rival if possible, such were the grand strategical misunderstandings. If nuclear war seemed out of the question, then the only avenue for confrontation and aggression would be conventional war. That neither side actually attacked the other with conventional weapons was sometimes said to be due to the fear that the any conflict would escalate to nuclear war. In this section I want to discuss a weapons acquisition programme that began in earnest in 1981, but which had its beginnings in a revision of US (conventional) military doctrine after the Vietnam War and the Yom Kippur War in 1973.

American involvement in Vietnam began in the early 1950s but increased considerably in 1961, with combat troops entering the country in 1965. The rationale was containment, which was often referred to as 'Domino Theory', the point being that if one regime, South Vietnam, fell to communism, other states in the region would fall like dominos. Again we have an example of grand strategy blinding the US to the truth, in this instance, a nationalist drive for unification of the country. The US withdrew from Vietnam in 1973, having lost the war, and the demoralised army turned its attention to problems in Europe that had never been properly addressed, namely how to deal with the threat of conventional war. In July of that year, the US army set up an institution called the Training and Doctrine Command or TRADOC, which eventually produced a new army field manual called FM 100-5. This document was much influenced by the experience of the Yom Kippur War which took place October 1973 between Israel on the one hand, and Egypt and Syria on the other.[17] The latter surprised Israel and nearly won, or nearly provoked Israel into using nuclear weapons, when Syrian tanks captured the Golan Heights for a short time. The relevance of this to the US problems of planning to deal with the Warsaw pact in Europe was that Israel was using American weapons when its opponents had Soviet systems and had been trained by Soviet advisors. So a real-life assessment of performance could be made, the Golan Heights resembled terrain in Western Europe, and finally the Israelis had beaten a much larger force and that would be what the US would be faced with in Europe.

The US concluded that a conventional war fought between technologically advanced opponents like the US and the Soviet Union would be short and very intense—like the Yom Kippur war but unlike the Vietnam War.[18] One reason was the immense firepower of modern weapons: the new field manual compared the performances of modern tanks, still the main weapon for conventional war, and found that once a target was 'acquired', the tank would hit it with 100% probability, as opposed to less that 10% in the Second World War (FM 100-5: 2.3). It also found that the front line Soviet tanks were probably better than the American ones. One of the 'big five' weapons systems that were called for after this review was a new tank, which eventually turned out be the massive M-1 Abrams main battle tank. The scenario for any European conventional war had the Warsaw Pact attacking NATO, so in this sense the US and its allies would be on the defensive, and hence the US military doctrine was defensive. In Chap. 4 we saw that fighting a defensive war does not imply never going on the offensive, and FM 100-5 devotes chapters to both defence and offence. The following passage is revealing

---

[17]For an overview, see Lee (2016: 426–436), and for more detail, see Bronfield (2007). See Prosch (1976) for an interview with an Israeli tank commander who defended the Golan Heights and is said to have prevented a Syrian breakout.

[18]The Vietnam War was an example of so-called asymmetric warfare, where one side is much more technologically advanced that the other.

While it is generally true that the outcome of combat derives from offensive operations, it may frequently be necessary, even advisable, to defend. Indeed, the defender has many advantages. Among these are the opportunity to know the terrain, to site and carefully emplace weapons and units, so as to minimize their vulnerabilities and maximize their capabilities, and reconnoiter and prepare the area for defense in depth. In fact, the defender has every advantage but one – he does not have the initiative. To gain the initiative, he must attack. Therefore, *attack is a vital part of all defensive operations*. (FM 100-5: 5.2, emphasis in original).

I will mention two ideas that emerged from this review of US doctrine and then come to the moral of the tale.

The US theatre strategy in Europe was christened Active Defense.[19] The aim of Active Defense was to hold the ground against a Soviet attack by destroying the attacker. Thus, after an initial (Soviet) attack, it was anticipated that combat would take the form of an Airland (Air Land) battle, a fight that involved tanks, artillery, mechanised infantry, helicopters and missiles systems on the part of the army, and all manner of aircraft on the part of the Air Force, all closely coordinated. The other side would fight the same battle, except that its aim was to break through by destroying the defender. One imagines that just by looking at what happened in an Airland battle and not knowing the circumstances, it would not be possible to tell who was the attacker and who was the defender. The Airland battle is thus kind of military operation, on a par with Blitzkrieg, with various tactics embedded therein. To fight Airland battles against numerically superior forces, the US army needed better equipment and more skilful soldiers. The former demand was to be met by the 'big five' weapons systems. In addition to the M1 Abrams tank, these were: the M-2 Bradley Infantry fighting vehicle, the UH-60 Blackhawk utility helicopter, the AH-64 Apache attack helicopter and the Patriot anti-aircraft defence system (Lee 2016: 435).[20]

The research needed to build these weapons was conducted after 1973, but full-scale production did not take place until Reagan's first term in office, which began in 1980. This was a low point in US-Soviet relations, marked by the infamous Star Wars speech and substantial spending on upgrading US nuclear forces. But it was followed by a remarkable thaw when Gorbachev became leader in Reagan's second term. Shortly thereafter, as everyone knows, the Soviet Union came to end, the Warsaw Pact dissolved and communism was no longer a threat to the US: the grand strategy that the US, and the Soviet Union, had pursued during the Cold War no longer had currency because all of a sudden there was no Cold War. The US now had the means to fight the Airland battle and defend Europe, in accordance with its conventional military doctrine, as well as having the means for nuclear deterrence, in accord with its nuclear doctrine. The rationale for all the

---

[19]I think it is easier to keep the 'levels of strategy' separated, and use Luttwak's nomenclature here. Bronfield refers to Active Defense as the US military doctrine in Europe, Bronfield (2007: 483).

[20]Descriptions of these systems can be found in Department of the Army (2008).

weapons research, both for conventional and nuclear weapons evaporated. But the 'Big 5' were all used in the two Gulf Wars, which were fought as Airland battles but against a completely outclassed opponent, so it seems that the US military doctrine about how to fight a conventional war survived the change in grand strategical context, which was so much the worse for Iraq and the middle east in general.[21]

6.6   Conclusion

To justify participation in a weapons research project, it is first necessary to estimate the harms caused by the weapons produced, to satisfy HC. If it is not possible even to do this, then the (much) more difficult tasks of deciding what harms have been prevented and then comparing the two will be impossible, and all hope of justifying the project will be abandoned. In this chapter we have seen how the context of weapons research can shift and change in ways which alter the deployments and uses of its products, making these impossible to predict, even for those experts who make policy, let alone the weapons researcher. Whether or not the weapons researcher is an expert privy to the most up to date information about the intended course of national policy is beside the point. It is not that were the weapons researcher to know everything about what the plans are for using the products of her research she could work out what harms they would cause: no one has this information.

References

Basalla, G. 1988. *The Evolution of Technology*. Cambridge: Cambridge University Press.
Benson, B. 2012. The Evolution of US Army Doctrine for Success in the 21st Century, *Military Review*. March-April, 2–12
Bishop, C. 2006. General editor. *The Encyclopaedia of Weapons*. San Diego: Thunder Bay Press.
Bronfield, S. 2007. Fighting Outnumbered: The Impact of the Yom Kippur War on the US Army. *Journal of Military History* 71 (2): 464–498.
Chase, K. 2003. *Firearms*. Cambridge: Cambridge University Press.
Chivers, C. 2010. *The Gun*. New York: Simon and Schuster.
Department of the Army. 2008. *US Army Weapons Systems*. New York: Skyhorse Publishing.
FM 100-5. 1976. http://cgsc.cdmhost.com/cdm/ref/collection/p4013coll9/id/42.
Ford, R. 2005. *The World's Great Machine Guns*. Leicester: Silverdale Books.
Forge, J. 2012. *Designed to Kill: The Case Against Weapons Research*. Dordrecht: Springer.

---

[21]Major Paquin of the US Armored Corps examined the question as to whether the US used Airland Battle in the first Gulf War and concludes that it did, see Paquin (1999). A revision of US doctrine after the war led to what was called Full Spectrum Operations, which included Active Defense as one element and made explicit the prevision for offensive operations, see Benson (2012).

Grant, N. 2015. *Mauser Military Rifles*. Oxford: Osprey.

Kahaner, L. 2007. *AK-47*. New Jersey: Wiley.

Lee, W. 2016. *Waging War*. Oxford: Oxford University Press.

MacLean, F. 1999. *The Field Men: The SS Officers Who Led the Einsatzcommandos*. Atglen Pa: Schiffer Military History.

McNeill, W. 1982. *The Pursuit of Power*. Chicago: Chicago University Press.

Paquin, R. 1999. *Desert Storm: Doctrinal AirLand Battle Success or 'The American Way of War'?*. School of Advanced Military Studies: Levenworth: Kansas.

Prosch, G. 1976. "Israeli Defence of the Golan". *Military Review* 59, October.

Shilin, V., and C. Cutshaw. 2000. *Legend and Reality of the AK*. Boulder, Colo.: Paladin.

Conclusion

To conclude, I will summarise my case against weapons research. The case proceeds in two stages: it is first of all established that weapons research is morally wrong, and then it is shown that it never possible to justify it. That two stages are required is a result of the particular moral system adopted, and I will begin with this observation.

The moral framework set up in Chap. 1 is non-consequentialist because it does not hold that the rightness and wrongness of an action is solely a function of the consequences of the action and that agents must seek to maximise 'the good' when they act, or minimise 'the bad'. In Chap. 1, on the other hand, I began with the Mill's Harm Principle and then by taking my lead from Feinberg, formulated

> HP: Do not harm others by invading, and so setting back, their rightful interests.

If all we know about an act is that it violates HP, then we can conclude that it is *prima facie* morally wrong. I also referred to Gert's system of common morality. This comprises ten moral rules, eight of which prohibit specific kinds of harming. HP is a generalisation (and slight reformulation) of Gert's rules which in turn are a (partial) specification of HP. It is convenient to have both ways of talking about harm in place. In particular, Gert's system is helpful when we turn to justified violations of the harm principle. I said in Chap. 1 that I believed most if not all consequentialist systems could also serve as a basis for the case against weapons research but that the argument would have a different form. In fact the two steps would be reversed.

It would undermine the institution of morality to take HP to hold absolutely and without exception, because every single time anyone harmed anyone else, caused them pain for instance, this would be morally wrong, and it would follow that moral judgements would be otiose. Everyone accepts that it can be necessary to cause some pain to prevent further pain and injury. Such an action may therefore be justified and the initial judgement of moral wrongdoing withdrawn. What I have referred to as paradigm instances are those involving a medical or dental procedure, intended to alleviate pain or prevent future pain, and such procedures are not of

J. Forge, *The Morality of Weapons Research*, SpringerBriefs in Ethics,
https://doi.org/10.1007/978-3-030-16860-5

course always instances of moral wrongdoing. Gert discusses the justification of violating moral rules at some length and argues convincingly that the only justification is in preventing or reducing further harms. It is not however acceptable to cause even a relatively small amount of harm to a few people to give a lot of pleasure to a greater number. Thus we have

> JP: The only justification for causing harm by violating the interests of moral subjects is the prevention of further harms.

JP and HP are the basic moral principles used to make the case against weapons research.[1]

The first stage in making the case against weapons research is to show that weapons research is morally wrong, and to do that we needed first to know something about the activity and come up with an description of it that we can accept as applying to all instances thereof. In Chap. 2 I began with two examples of weapons research. The first of these, the Manhattan Project, showed how scientific theory could guide and inform weapons design, and were we to restrict attention to examples of this kind, we would conclude that weapons research is a species of applied science. But the second example, that of the torsion catapult, provides an alternative view, that research in the sense of careful experiment and testing can come up with new weapons without the underlying theory being available. The reasons in favour of adopting this second position are all the more persuasive because the Greek engineers left detailed designs about how to build a whole range of torsion artillery, complete with mathematical equations for calculating the key parameters. I concluded that we should date weapons research from at least 400 BCE and, on the basis of a quick examination of some even older weapons, probably earlier still. This led to

> Weapons research is research carried out with the intention of designing new weapons or improving the design of existing weapons or designing or improving the means for carrying out activities associated with the use of weapons.

The last section of the chapter introduced the notion of the context of weapons research.

Research is not in itself normally a harmful activity, so if HP is to have any bearing on weapons research, some connection must be established between weapons research and harming. This is done is Chap. 3, where I argued that weapons are the means to harm, that it is morally wrong to provide the means to harm and that weapons research provides the means to harm. I introduced a taxonomy of purposes for artefacts, which distinguished primary, derivative and secondary categories. The primary purpose is what the artefact is intended to do,

---

[1]To arrive at a judgement of moral rightness or wrongness on the basis on a consequentialist principle it is necessary first to determine the nature of the consequences. On the basis of principle CP—see Chap. 1—for example it would be necessary to see if the act maximally promotes the interests of others. Then a conclusion about the act could be made and my two steps are therefore reversed.

what function it is intended to perform. I believe that it is possible to identify a primary purpose for almost any artefact, included those 'generic' artefacts that seem highly versatile and have many uses. For weapons, however, it is relatively easy to show that they are primarily the means to harm, and that deterring harm, for instance, is a derivative purpose, which presupposes that they are the means to harm but not conversely. This leads to

PP1: The primary purpose of weapons research is to design new ways to harm

I claim that weapons researchers are therefore responsible for the primary purpose of creating new or improved ways to harm, for this is what they intend when they undertake weapons research. They may also be responsible for certain derivative uses, though that depends on the example in question. Designers are not normally responsible for secondary purposes, which are fortuitous.

In the last section of Chap. 3 I maintained that it is wrong to provide the means to harm. This was done with the aid of an example of a bomb maker for a terrorist organisation. The example was deliberately chosen because such organisations are illegal and (nearly) everyone condemns them, especially when they use bombs. It is thus hard to deny that the weapons designer D who makes the bombs does the wrong the thing. However, when it comes to moral wrongdoing, there is no relevant difference between D when she draws up designs for bombs without knowing when or if they will be made or used and the 'legitimate' weapons designers who works for a state-run organisation, because (all) the uses of these weapons are not known in advance either. The atomic bombs produced in the Manhattan Project killed many more innocent people than all the terrorist activity since. We can therefore assert

MP. If it is morally wrong to harm, then it is morally wrong to provide the means to harm.

and by HP infer

WRMR: Weapons research is morally wrong.

It does *not*, however, follow that the weapons researcher is to be held responsible and called to account for all the harms caused by the weapons produced, although this is possible and it depends on the circumstances. Weapons research is wrong because it introduces new ways of harming into the world and therefore risks those new ways being used to cause harm. Weapons researchers are guilty of wrongdoing because they intentionally design new ways to harm.

Defence is the rationale for everything to do with weapons acquisition, including weapons research, by almost everyone, as has been the case in the past. The aim of Chap. 4 was to show that this apparently reasonable viewpoint does not stand up to scrutiny. The problem is that it seems that weapons which can be used to defend against aggression and so prevent harm, in accordance with JP, can also be used for aggression. But perhaps there is a class of weapons that can only be used to prevent aggression, and in which case there would be some weapons research that was not morally wrong. This suggestion was examined and found wanting: there are no such weapons, and for two reasons. In the first place it appears that there are no weapons that can only be used to preserve rather than attack an asset, assuming that preserving assets is the essence of defence. I conjectured that the possible use of a weapon in a defensive role was a matter of what else was available, how the weapon could be function in the context of the available military technologies, and the evidence seemed to support this viewpoint. In the second place, however, when considering the levels of strategy on which the scale and nature of fighting and planning in warfare can be differentiated, we saw that defending assets was a necessary part of offensive war, a proposition confirmed in Chap. 6.

The first conclusion of Chap. 4 is that how weapons are used is a matter of the context in which they are employed, not on their 'nature'. The second conclusion concerns deterrence. In the last section of the chapter, we saw that deterrence is a relationship between states where one thinks that the other has certain aggressive designs and seeks to prevent those being carried out by threatening to impose unacceptable costs on the aggressor, and acquires weapons to reinforce its stance. But relations between states can change, as they did at the end of the Cold War, and weapons obtained for the ends of deterrence will no longer be needed for that purpose, but they are still the means to harm—just as is the case for defence, there are no inherently deterrent weapons. The third and final conclusion of Chap. 4 is that there can be no excuse or justification for weapons research that appeals to the inherent nature of the weapon in question, divorced from the context of its use. Or, to put the matter another way, there can be no ahistorical justification for weapons research.

If there can be no ahistorical justification for weapons research, then all attempts at justification must be *historical*, they must refer to the situations and circumstances in which weapons are actually used, and they must be such that JP is satisfied. Chapters 5 and 6 show that this demand cannot be met and hence that weapons research cannot be justified. In the first section of Chap. 5 I said that even for those paradigm cases of justified harming, namely medical procedures, it can be difficult to estimate the harms caused and harms prevented, the point being that if it is not easy to do so in these well-understood and circumscribed situations, how much harder must it be for estimating the harms caused and prevented by the products of weapons research. Indeed, even the easier task of estimating the harms caused looks impossible. As a way of trying to get round this problem, I considered two possibilities: that weapons research conducted by democratic states is justified because such states are never aggressors and only fight wars that are defensive and hence just, that weapons research conducted in the prosecution of a just war (regardless who is involved) is justified. Neither of these manoeuvres is successful. There are many examples of democratic states waging aggressive war, and the stronger they are in terms of military muscle, the more they tend to do so. As for just war, I have argued that the unknowable costs consequent on weapons research is incompatible with the *ad bellum* proportionality condition of JWT, and this means that, far from being justified, weapons research is proscribed by the theory.

In Chap. 6 I gave examples which show that it is practically impossible to work out the harms caused by weapons because it is impossible to know how they will be used in the future. While this may well seem obvious, I gave some reasons why it is true with reference to the contexts in which weapons research is conducted and in which weapons are used. The idea of context was discussed in the first section, and having mentioned and discussed the notion of the grand strategy of a state and the corresponding military doctrine, I suggested that we could make the idea a little more formal by agreeing that grand strategy is the most important element. Thus as grand strategy changes so does the context. The three examples all showed how the products of weapons research persist through changes in context and military doctrine, matters that no one could anticipate.

Final Words

In the Introduction to this book, I reflected on the possibility of weapons research coming to a halt if I could successfully make the case against it, show that it is morally wrong. I suggested that the chances of that happening were remote indeed —but hoped the book would be an interesting contribution to philosophy, which would be enough for me! I then made three comments, including noting that if moral progress is possible, someone needs to take the first step. But what would happen if states began to give up weapons research? Those with vested interests, and those seduced by the standard rationale, will deny that this is a good idea, or even possible. States who do not modernise their defence forces will become weaker and prey to pragmatic states who continue to invest in weapons research.

But that need not be true. Consider again states A and B, the erstwhile rivals: instead of B embarking on an intense round of weapons research in response to perceived threats from A, it phases out all its weapons research. What would A's reaction be, wait until it has clear military superiority and then coerce B to do its bidding? Why should this happen? Could not A now see that B is not a threat to its security and decide to slow down its own weapons research programmes, and phase them out as well when it sees how much money it has saved? We have here a reversal of Jervis' security dilemma: states feel more and more secure so continue to reduce their armed forces, instead of an arms race, we would have a 'disarmaments race'. Why could not this happen? If it did, then countries could get on with problems that really do threaten their security, like climate change and poverty, things which really affect people and cause them harm. States could cooperate and take steps to reduce these harms rather than spending huge sums of money producing new ways to harm. Surely that makes sense?

Index

© The Author(s), under exclusive licence to Springer Nature Switzerland AG 2019   101
J. Forge, *The Morality of Weapons Research*, SpringerBriefs in Ethics,
https://doi.org/10.1007/978-3-030-16860-5